FOREWORD

Water is essential for life and for economic activity. Yet in large parts of the world, many people and enterprises do not have access to safe drinking water or sanitation services.

In the New Independent States of the former Soviet Union (NIS), problems of access to water services are rooted in the history of that region. Ambitious investment programmes led to the development of extensive networks of water infrastucture in urban areas. However, these networks were poorly designed and constructed with inferior materials, and they have not been adequately maintained. As a result, urban infrastructure is on the point of collapse in many NIS cities, with potentially calamitous consequences for human health and economic activity. Considerable efforts by International Financial Institutions (IFIs) and donors to mitigate these problems have achieved only modest results.

The Almaty Ministerial Consultation held in Almaty, 16-17 October, 2000 was convened to address these pressing problems. It brought together Economic/Finance and Environment Ministers from the NIS with ministers from several OECD countries. Senior representatives from other OECD countries, IFIs and international organisations also participated. Non-governmental organisations held three regional conferences and provided a constructive input to the debate. The private sector, which has begun to participate in the management and operation of water services in some NIS cities, contributed their experience.

The main achievement of the meeting was to establish a consensus among the major stakeholders on the critical situation which exists in the NIS, and to agree on some of the key steps which need to be taken in order to address it. Ministers endorsed a set of Guiding Principles for Reform of the Urban Water Sector in the NIS which will provide the framework for implementing a concerted action programme involving the major stakeholders.

The action programme will be implemented through the Task Force for the Implementation of the Environmental Action Programme for Central and Eastern Europe (EAP Task Force), which also provided the framework for preparing the Consultation. The EAP Task Force was established with its Secretariat in OECD's Environment Directorate, Non-Member Countries Division, following a decision taken at the 1993 "Environment for Europe" Ministerial Conference in Lucerne. The work forms part of OECD's Programme of Co-operation with Non-Members.

This volume presents the main papers from the Consultation and is published under the authority of the OECD Secretary-General. The views expressed do not necessarily reflect those of OECD or its Member countries.

Joke Waller-Hunter Director Environment Directorate	Éric Burgeat Director Centre for Co-operation with Non-Members

© OECD 2001

Water Management in the Newly Independent States

PROCEEDINGS
BETWEEN
AND

16-17

ORGANISATION FOR ECONOMIC CO-OPERATION AND DEVELOPMENT

Pursuant to Article 1 of the Convention signed in Paris on 14th December 1960, and which came into force on 30th September 1961, the Organisation for Economic Co-operation and Development (OECD) shall promote policies designed:

- to achieve the highest sustainable economic growth and employment and a rising standard of living in Member countries, while maintaining financial stability, and thus to contribute to the development of the world economy;
- to contribute to sound economic expansion in Member as well as non-member countries in the process of economic development; and
- to contribute to the expansion of world trade on a multilateral, non-discriminatory basis in accordance with international obligations.

The original Member countries of the OECD are Austria, Belgium, Canada, Denmark, France, Germany, Greece, Iceland, Ireland, Italy, Luxembourg, the Netherlands, Norway, Portugal, Spain, Sweden, Switzerland, Turkey, the United Kingdom and the United States. The following countries became Members subsequently through accession at the dates indicated hereafter: Japan (28th April 1964), Finland (28th January 1969), Australia (7th June 1971), New Zealand (29th May 1973), Mexico (18th May 1994), the Czech Republic (21st December 1995), Hungary (7th May 1996), Poland (22nd November 1996), Korea (12th December 1996) and the Slovak Republic (14th December 2000). The Commission of the European Communities takes part in the work of the OECD (Article 13 of the OECD Convention).

OECD CENTRE FOR CO-OPERATION WITH NON-MEMBERS

The OECD Centre for Co-operation with Non-Members (CCNM) promotes and co-ordinates OECD's policy dialogue and co-operation with economies outside the OECD area. The OECD currently maintains policy co-operation with approximately 70 non-Member economies.

The essence of CCNM co-operative programmes with non-Members is to make the rich and varied assets of the OECD available beyond its current Membership to interested non-Members. For example, the OECD's unique co-operative working methods that have been developed over many years; a stock of best practices across all areas of public policy experiences among Members; on-going policy dialogue among senior representatives from capitals, reinforced by reciprocal peer pressure; and the capacity to address interdisciplinary issues. All of this is supported by a rich historical database and strong analytical capacity within the Secretariat. Likewise, Member countries benefit from the exchange of experience with experts and officials from non-Member economies.

The CCNM's programmes cover the major policy areas of OECD expertise that are of mutual interest to non-Members. These include: economic monitoring, structural adjustment through sectoral policies, trade policy, international investment, financial sector reform, international taxation, environment, agriculture, labour market, education and social policy, as well as innovation and technological policy development.

© OECD 2001
Permission to reproduce a portion of this work for non-commercial purposes or classroom use should be obtained through the Centre français d'exploitation du droit de copie (CFC), 20, rue des Grands-Augustins, 75006 Paris, France, tel. (33-1) 44 07 47 70, fax (33-1) 46 34 67 19, for every country except the United States. In the United States permission should be obtained through the Copyright Clearance Center, Customer Service, (508)750-8400, 222 Rosewood Drive, Danvers, MA 01923 USA, or CCC Online: *www.copyright.com*. All other applications for permission to reproduce or translate all or part of this book should be made to OECD Publications, 2, rue André-Pascal, 75775 Paris Cedex 16, France.

TABLE OF CONTENTS

List of abbreviations and country groups	9
Introduction	11
Chapter 1. **Joint Conclusions of Almaty Ministerial Consultation**	15
Water in the NIS	15
Integrating Economic and Environmental Decision-making	16
Finance Strategies	16
Reform of the Urban Water Supply and Sanitation Sector	17
Private Sector Participation	18
Public Participation and Consumer Protection	18
Next Steps	18
Chapter 2. **Background Paper on Valuing Environmental Benefits and Damages in the NIS: Opportunities to Integrate Environmental Concerns into Policy and Investment Decisions**	21
I. Introduction	21
II. Economic valuation as a tool for integrating environmental concerns in decision-making	22
III. Recent practices in using environmental valuation in decision-making in the NIS	27
IV. Applications of environmental valuation in decisions concerning water pollution control and urban water supply	32
V. Pilot studies under EAP Task Force Program	37
VI. Conclusions and recommendations	39
Notes	41
Appendix 1. Summary of environmental valuation studies conducted in NIS	42
Appendix 2. Some pilot projects proposed by the NIS for further application of modern valuation methods	45
Chapter 3. **Background Paper of Financing Strategies for the Urban Water Sector in the Nis**	47
1. Introduction and key policy implications	47
2. Overview of financing water and sanitation infrastructure in the NIS	48
3. Strategic planning and programming in the water and sanitation sector	51
4. Results of the first financing strategies in the NIS	55
Chapter 4 **Background Paper on Reform of Urban Water Supply and Sanitation Services**	73
Introduction	73
Notes	88
Appendix 1. Recommendations for Successful Involvement of the Private Sector in the Provision of Water Supply and Sanitation Services	89
Chapter 5. **Private Sector Participation in Urban Water Supply and Wastewater Financing and Management: an Opportunity for Increased Financing and Improved Efficiency**	91
1. Introduction	91
2. Developing countries and economies in transition: emerging models for private investment	93
3. Lessons learned about private investment in urban water and waste water services	100
4. Steps governments and other actors can take to increase private investment in urban water and waste water services	104

© OECD 2001

Chapter 6. **Water Supply and Sanitation the Private Sector's View on Risks and Opportunities** 111
General Introduction 111
Investment in Water Supply and Sanitation 112
Potential Risks from the Private Sector's Viewpoint 113
Benefits to the Public Sector of Public – Private – Partnerships 115
Benefits to the Private Sector of Public – Private – Partnerships 115
Key Conditions for Successful Contracts 115
Involving stakeholders 116
The Transition 117
Conclusions 117
Notes 118

Chapter 7. **Position Paper of the Non-governmental Organisations on the Main Discussion Issues of the Almaty Ministerial Consultations** 119
1. Integrating economic and environmental decision-making 119
2. Reform of the urban water supply and sanitation sector 120

Annex I. Guiding principles for reform of the urban water supply and sanitation sector in the NIS 123
Annex II. List of participants 129

List of Boxes

1. When should environmental expenditures be made? 23
2. Fighting infectious diseases in Orlovsky District, Rostov Oblast, Russia 24
3. Categories of economic value 24
4. Valuing the health and safety effects of environmental degradation: physical linkage methods 25
5. Valuing non-market goods on the basis of stated preference: Contingent Valuation Method (CVM) 26
6. Determining the demand for environmental goods and services: revealed preference methods 26
7. Comparison of estimates of environmental damage costs 28
8. Using valuation for environmental priority setting at the national level – the Moldova NEAP 29
9. Using valuation for prioritising investment at the plant level – Severstal Iron and Steel Works 30
10. Guidelines for economic valuation studies: 32
11. Regulatory impact analysis of lead in drinking water 33
12. Benefit cost analysis of a proposed investment program for nutrient reduction in the Danube River and the Black Sea 35
13. Contingent valuation of municipal services in Iasi, Romania 36
14. Clean Water Program in the city of Rostov-on-Don, Russia 37
15. An Assessment of the environmental benefits of alternative water supply interventions – pilot study in Astana, Kazakhstan 38

1. Water supply losses in the NIS 74
2. Possible Structure for a Water Sector Reform Group (WSRG) at national level 77
3. Recommendations water sector responsibilities for national authorities 79
4. Public participation and protection of the poor and consumers 81

1. Strengths and weaknesses of service contracts 94
2. Strengths and weaknesses of BOT arrangements 95
3. Strengths and weaknesses of concession contracts 96
4. Strengths and weaknesses of joint ventures 97
5. Strengths and weaknesses of full divestiture 98
6. Strengths and weaknesses of unregulated private provision 98

List of Tables

1. Suggested applications of environmental valuation at different levels of decision-making 27

1. Shares of general government tax revenue in GDP in 1998 49
2. Annual expenditure needs to properly operate and maintain existing water and sanitation infrastructure at 1999 levels of service and available finance under business as usual scenario in Georgia 58

3. Annual expenditure needs to properly operate and maintain existing water and sanitation infrastructure at 1999 levels of service and available finance under business as usual scenario in Moldova 62
4. Policies and measures sufficient to close the financing gap and restore the present level of service within less than 20 years 66
5. Annual expenditure needs to properly operate and maintain existing water and sanitation infrastructure at 1999 levels of service and available finance under business as usual scenario in Novgorod 68
1. Some WTP results from the NIS city affordability studies examined 81

List of Figures

1. Illustration of general government balances in CEE countries and the NIS 50
2. Simplified structure of the model used to development of financing strategies 54
3. Sectoral policy cycle and the role for financing strategies 54
4a. Operations, maintenance, rehabilitation and extension of infrastructure 55
4b. Financing gap for the scenario of partial rehabilitation of Tbilisi system 59
5. Financing gap for the scenario of rehabilitation of Tbilisi and the Black Sea coast WWTPs by 2020 60
6. Maintenance backlog for the scenario of rehabilitation of Tbilisi systems and the Black Sea coast WWWTPs 61
7. Partial impact of increasing user charges up to affordability level on the backlog of maintenance 64
8. Partial impact of additional public finance on the backlog of maintenance 65
9. Partial impact of US$62 million concessional loan on the backlog of maintenance 65
10. Combined impact of all additional finance measures on the backlog of maintenance 66
11. Financing gap under the baseline and the scenario of achieving targets of improvement in quality and level of services (in US$ million) 70
12. Backlog of maintenance under the baseline and the scenario of achieving targets of improvement in quality and level of services (in US$ million) 70
1. Typical NIS Arrangement for urban water services delivery 76
1. Regional share of FDI in the water sector 112
2. EBRD transition indicators and investment climate survey scores 114

LIST OF ABBREVIATIONS AND COUNTRY GROUPS

BOD/l	Biological Oxygen Demand per Litter
CEE	Central and Eastern Europe
CIS	Commonwealth of Independent States
COWI	A Danish Consulting Company
CPPI	Centre for Preparation and Implementation of International Projects on Technical Assistance
DANCEE	Danish Co-operation for Environment in Eastern Europe
DEPA	Danish Environmental Protection Agency
EBRD	European Bank for Reconstruction and Development
EFS	Environmental Finance Strategy
EMP	Environmental Management Programme
EU	European Union
GDP	Gross Domestic Product
GEF	Global Environmental Facility
GOST	State Standard (Gosudarstvennyj Standart)
HIID	Harvard Institute for International Development
IDA	International Development Association
IFC	International Financial Corporation
IQ	Intellectual Quotient
MAC	Maximum Allowable Concentration
MCL	Maximum Concentration Level
NEAP	National Environmental Action Programme
NIS	New Independent States of the former Soviet Union
O&M	Operation and Maintenance
PPC	Project Preparation Committee
PPP	Polluter Pays Principles
RAMSAR	Convention on Wetlands of International Importance especially as Waterfowl Habitat, 1971
REC	Regional Environmental Centre
Rio+10	World Summit on Sustainable Development
SEE	South East Europe
UN ECE	United Nations Economic Commission for Europe
UNDP	United Nations Development Programme
UNEP	United Nations Environment Programme
UNICEF	United Nations Children's Fund
USAID	United States Agency for International Development
USEPA	United States Environmental Protection Agency
WBCSD	World Business Council for Sustainable Development
WHO	World Health Organisation
WWT	Waste Water Treatment

OECD countries

The original Member countries of the OECD are Austria, Belgium, Canada, Denmark, France, Germany, Greece, Iceland, Ireland, Italy, Luxembourg, the Netherlands, Norway, Portugal, Spain, Sweden, Switzerland, Turkey, the United Kingdom and the United States. The following countries became Members subsequently through accession at the dates indicated hereafter: Japan (28th April 1964); Finland (28th January 1969); Australia (7th June 1971); New Zealand (29th May 1973); Mexico (18th May 1994); the Czech Republic (21st December 1995); Hungary (7th May 1996); Poland (22nd November 1996); the Republic of Korea (12th December 1996); and the Slovak Republic (14th December 2000). The Commission of the European Communities takes part in the work of the OECD (Article 13 of the OECD Convention).

Central and Eastern European countries

Albania, Bulgaria, Croatia, Czech Republic, Estonia, Hungary, Latvia, Lithuania, Macedonia (Former Yugoslav Republic of), Poland, Romania, Slovenia.

(Note: The report does not cover Bosnia and Herzegovina or Yugoslavia).

The New Independent States of the former Soviet Union (NIS)

Armenia, Azerbaijan, Belarus, Georgia, Kazakhstan, Kyrgyz Republic, Moldova, Russian Federation, Tajikistan, Turkmenistan, Ukraine, Uzbekistan.

INTRODUCTION

At the 1998 "Environment for Europe" Conference in Aarhus, Denmark, Ministers asked the EAP Task Force to give priority to addressing the pressing environmental problems in the New Independent States of the former Soviet Union (NIS).* The underlying strategy of the EAP Task Force in taking up this challenge has been to promote better integration of economic and environmental decision making; and the priority sector for attention has been water.

The Task Force considered that it was essential to strengthen political support for its work in the NIS, and to this end, agreed to convene a meeting of NIS Economic/Finance and Environment ministers together with other stakeholders. It was agreed that the main focus of the meeting would be on the problems of the urban water sector, but that this issue should be treated in the broader framework of integrating economic and environmental decision-making.

The meeting proved very timely. As the papers in this volume demonstrate, urban water supply and sanitation services are on the point of collapse in many NIS. The policy and institutional reforms required to put the sector on a sound and sustainable basis, and to create the conditions for much needed investments, have generally not been implemented. As a result, the considerable efforts of International Financial Institutions and donors over the last 10 years have had only a modest impact.

The consequences of failing to reform the urban water supply and sanitation sector are now clear:

- Water usage is excessive by international standards, with high levels of wastage by consumers and in distribution networks.
- The supply of water is unreliable and of low quality, with the poor particularly affected.
- Water and wastewater treatment is increasingly ineffective.
- In some countries there have been increases in water-borne diseases, adverse impacts on industrial and agricultural productivity and an impairment of the ecological functions of aquatic systems.

Probably the most important achievement of the Almaty Consultation was establishing a consensus among the major stakeholders on the critical situation of urban water infrastructure and its management in the NIS, and on some of the key steps which need to be taken to address it. In addition to NIS Ministers of Economics/Finance and Environment, Ministers and senior representatives from several OECD countries participated, as well as senior officials from International Financial Institutions and International Organisations active in the region. Non-governmental organisations and private sector representatives also shared their experience.

---

* The Environment for Europe process was launched at a Ministerial meeting in Dobris, Czechoslovakia in 1991. Subsequently Ministerial meetings were held in Lucerne, Switzerland (1993); Sofia, Bulgaria (1995); Aarhus, Denmark (1998). The next is scheduled for 2003 in Kiev, Ukraine. At the Lucerne Conference in 1993, Ministers endorsed the Environmental Action Programme for Central and Eastern Europe (EAP). They also established two bodies to facilitate its implementation: the EAP Task Force, with its Secretariat at OECD, was established to promote environmental policy and institutional reform. The Project Preparation Committee (PPC), with its Secretariat at the European Bank for Reconstruction and Development, aims to accelerate environmental investments in the region. For further information on these and related activities, see OECD (1999) "Environment in the Transition to a Market Economy: Progress in Central and Eastern Europe and the New Independent States".

© OECD 2001

The conclusions of the meeting are presented in Chapter 1. They identify the main findings and recommend how some of the key challenges could be addressed. Participants endorsed Guiding Principles for Reform of the Urban Water Sector in the NIS as a basis for further action and called on all stakeholders to support their implementation. The Guiding Principles are set out in Annex I. Participants also requested the EAP Task Force to develop a focussed programme of work to facilitate implementation of the Guiding Principles. Progress will be reviewed at the next "Environment for Europe" Ministerial Conference, to be held in Kiev, Ukraine in 2003, and at a major conference of stakeholders to be held no later than 2005.

The subsequent chapters present the main papers prepared for the Consultation together with the views of the non-governmental participants.

Chapter 2 examines the integration of economic and environmental decision-making in the NIS. A fundamental problem in the NIS is the perception that environmental protection can only be achieved at the expense of economic development. In the difficult circumstances that most NIS now find themselves in, this view is often used to legitimise "pollute now, clean up later" strategies. However, such strategies have been discredited in OECD countries and many emerging economies. They ignore the significant economic costs which can result from poor environmental policies and the many opportunities that exist for designing economic and environmental policies in ways that are mutually supportive. The Chapter emphasises how new methods of project and programme appraisal can support the design of more effective economic and environmental policies, contribute to the reform of public decision-making and help to ensure the sustainability of economic activities. To this end, the World Bank is implementing a capacity building programme within the framework of the EAP Task Force to support the greater use of such methods in decision-making in the NIS.

Cost-benefit analysis and similar techniques can help in establishing policy priorities and targets. However, once targets have been established, it is crucial to ensure that they are achieved in the most cost-effective manner. Chapter 3 examines how finance strategies for the urban water supply and sanitation sector can assist this task. With Danish support, the EAP Task Force developed a tool for preparing such strategies and worked with several countries – Georgia, Kazakhstan, Moldova and Ukraine – and two regions of the Russian Federation – Novgorod and Pskov – to apply it. This work showed that even to operate and maintain water infrastructure at current performance levels would require implementation of demanding packages of policy and institutional reforms. The time that will be required to meet standards enjoyed in the OECD countries will probably be measured in decades rather than years. The finance strategies indicate that increased user charges are the only realistic means of covering operation and maintenance charges; however adequate provision would also need to be made to ensure that poor and vulnerable groups have access to water services. Public budgets will have a crucial role to play in financing capital investments and providing social protection. Finally, the finance strategies show that while projects involving international financial institutions could have an important demonstration and catalytic role, their numbers will be limited by countries' limited borrowing capacity.

Investments are a necessary but not sufficient condition for reversing the deterioration of urban water services in the NIS: fundamental policy and institutional reforms are also required to put the sector on a sound and sustainable footing. Without such reforms, increased financing resources would reinforce the inefficiencies that currently exist, weaken the need for change and increase the costs of remediation. Chapter 4 examines these issues and provides the rationale for the Guiding Principles for Reform of the Urban Water Supply and Sanitation Sector in the NIS. The Guiding Principles emphasise the need for decentralisation of decision-making; a clear legal and administrative framework which delineates the roles of local governments and water utilities; establishing the sector on a financially sustainable basis; and involving the public in the reform process. The EAP Task Force will build on the lessons learned from reform efforts over the last 10 years and focus further work on those elements of the Guiding Principles which are of most strategic importance in promoting urban water sector reform throughout the NIS.

One issue that provoked different views in Almaty was the possible role of the private sector. Some participants considered that private sector participation in the management or operation of water utilities, particularly in large cities, was the most effective – perhaps the only – way to drive reform. Other participants pointed to the extensive positive experience with publicly managed utilities in many other countries. Chapter 5 reviews global experience with private sector participation in urban water sector reform, and Chapter 6 presents a private sector perspective on public-private partnerships in the operation of water utilities in the NIS. The chapters emphasise that private sector participation is not equivalent to privatisation; the private sector can be involved in the management or operation of water utilities in a variety of ways. Moreover, all forms of private sector participation require a strong regulatory framework to provide appropriate safeguards for both consumers and private utility operators. Most participants agreed on the need to go beyond "public v. private" debates and to identify the arrangements which could best ensure efficient and effective delivery of urban water services. The EAP Task Force will conduct further analysis of those issues, particularly the experience in central and eastern Europe, and promote further dialogue among key stakeholders.

Chapter 7 presents the views of non-governmental organisations (NGOs). The paper was developed on the basis of three regional conferences, involving 111 NGOs from all the NIS. The paper reflects the dissatisfaction of consumers with the poor quality of urban water services. It supports the urgent need for reform and identifies ways in which NGOs and the public can contribute constructively. This will be vital since improvements in water services and quality will almost certainly lag increases in user charges. The NGOs also emphasised that reformed water utility should function in a fully transparent and accountable manner; the non-transparent way in which utilities currently operate creates suspicion that utility reform, particularly higher user charges, will benefit managers rather than consumers. They also stressed that special provisions would need to be made to cushion poor and vulnerable groups from increased user charges. This could best be achieved by targeted subsidies rather than the current approach which treats water as a free good and thereby undermines incentives for efficiency and the financial viability of the sector. Paradoxically, it is frequently the poor and vulnerable who suffer most from the inadequacies of the current system; these groups therefore also stand to gain from reforms which explicitly address their needs.

Chapter 1

JOINT CONCLUSIONS OF ALMATY MINISTERIAL CONSULTATION

We[*] met in Almaty, Kazakhstan, 16-17 October 2000 to discuss how to better integrate economic and environmental decision-making and, thereby, to contribute to sustainable development in the New Independent States which emerged on the territory of the former Soviet Union (NIS). Our discussion focussed on the serious problems in the urban water sector in the NIS.

We are grateful to the government of the Republic of Kazakhstan for its leadership in organising this event and for the hospitality shown to participants during the conference.

Water in the NIS

Water is vital for life and the health of people and ecosystems, and it is a basic requirement for sustainable economic development. In the NIS, there are many pressing problems in the water sector related to both water quality and quantity. These problems vary across the region, they occur in both rural and urban areas, and in some regions they have important transboundary dimensions. We recognise that integrated approaches, organised around river basins, provide the most effective means of managing water resources. Effective management of water, nationally and across boundaries, is also important for peace and security within the region.

Water and sanitation services in urban areas are an important priority in the NIS, because of the direct human health impacts they have on large numbers of people concentrated in towns and cities. In urban areas in the NIS:

- Water usage is excessive by international standards, with high levels of wastage by consumers and in distribution networks;
- The supply of water is unreliable and of low quality, with the poor particularly affected;
- Water and wastewater treatment is increasingly ineffective;
- Insufficient attention has been given to waste water treatment relative to water supply;
- There are increases in water-borne diseases, adverse impacts on industrial and agricultural productivity, and an impairment of the ecological functions of aquatic eco-systems.

The NIS share with other regions of the world many of the same problems in ensuring adequate provision of urban water and sanitation services. However, the problems which the NIS face also have a number of distinctive features. Generally the percentage of the population served by piped water and connected to sewerage and waste water treatment plants in urban areas is higher than in countries of a similar income level. The first challenge in the NIS therefore is to maintain the quality of existing infrastructure and then to improve it.

---

[*] Armenia, Azerbaijan, Belarus, Czech Republic, Denmark, Finland, France, Georgia, Germany, Kazakhstan, Republic of Kyrgyzstan, Republic of Moldova, the Netherlands, Norway, Russian Federation, Switzerland, Tajikistan, Turkey, Turkmenistan, Ukraine, United Kingdom, United States, Uzbekistan; European Commission, OECD, EAP Task Force, World Bank, IFC, EBRD, PPC, UN ECE, UNEP, UNDP, UNICEF, WHO, CIS Executive Committee; private sector (WBCSD); Regional Environmental Centres (RECs for Central and Eastern Europe, Central Asia, Moldova, Russian Federation); non-governmental organisations (European ECO-Forum).

Although some important initiatives have been taken to meet this challenge they have not been sufficient. The fundamental reforms required to put the urban water and sanitation sector on a sound and sustainable basis have not been taken. As a result, the water supply and sanitation services in many NIS are now in critical condition and deteriorating. Without urgent action by governments of the NIS, the quality of the services will continue to worsen and, in some countries, may even collapse, with serious consequences for the health and well-being of the populations and their environment.

Integrating Economic and Environmental Decision-making

Addressing the crisis in the urban water sector, and environmental problems more generally, requires a better integration of economic and environmental decision-making. Governments should take into consideration environmental costs and benefits, in addition to other economic costs and benefits, and identify least-cost solutions. However, there are many opportunities to achieve economic and environmental objectives at the same time ("win-win" policies). We will strive to develop such policies, and we reject "pollute now clean-up later" strategies as being ineffective and more costly than pollution prevention strategies.

We recognise the need to better integrate economic and environmental decision-making as part of the broader reform of decision-making by governments. New methods for project and programme appraisal should be employed in the NIS so that the costs and benefits of improved water supply and sanitation can be assessed against other areas competing for public finance. To this end, NIS governments with donor support, should strengthen the information base, train policy advisors and reform decision-making procedures. Integration between government agencies responsible for environment, economy and finance should be a principle of government policy. We welcome the project which the World Bank is supporting within the EAP Task Force framework to build capacity in these areas, and call upon Task Force members to support this and similar initiatives.

Greater public awareness of environmental and economic links will facilitate sectoral integration. To this end, information campaigns about the links between environmental quality, public health and economic development will be instrumental. NGOs should be invited to actively participate in the organisation of such campaigns.

Finance Strategies

Deterioration of water and sanitation services in the NIS is closely linked to the sharp declines in the allocation of public finance to this sector since 1991. Greater financial resources are needed to solve the problems of the water sector in the NIS. However, we recognise that existing financial resources should be used more effectively to achieve greater environmental benefits, to ensure better services and as a lever to mobilise additional financing. We welcome the development, with Danish support, of an environmental finance methodology. Results emerging from the application of this methodology to the water sector in Georgia, Kazakhstan, Moldova, and two regions of the Russian Federation are helping to establish more realistic targets for the urban water sector in the NIS and to identify some of the key elements of finance strategies required to achieve them.

National and local budgets have an essential role in the short and medium term in financing rehabilitation and capital investments, in providing social protection and in facilitating access to credit. Scarce public funds and donor grants should be concentrated on a few projects to ensure that urgently needed rehabilitation is carried out, and that the deterioration of water networks is arrested. Preparation and implementation of financial arrangements should be conducted transparently in order to enhance efficiency and effectiveness. IFI projects will continue to have an important demonstration and catalytic function, but their numbers and size will be limited by countries' borrowing capacity. In order to ensure the sustainability of this process, participation of national firms in the preparation and implementation of projects that receive full or partial foreign finance, should be enhanced. Appropriate consideration should be given to the level of national participation and national capacity-building when launching tenders to ensure projects have sustainable results.

Continuation of the current combination of ambitious targets for the level and quality of water services and financing arrangements, which involve low user charges and limited access to funds for capital repairs, is not sustainable. This combination will result in a further deterioration of the level and quality of services. User charges are the only feasible long-term source of finance for operation and maintenance expenditure. At the same time, we recognise that increases in user charges must take full account of what people can afford. Existing subsidy schemes should be replaced by targeted support for poor and vulnerable groups, as a part of a strategy for service provision that has been developed through a participatory process.

Economic/Finance, Environment and other relevant Ministries in the NIS will continue to co-operate in developing such strategies and in integrating the results into multi-year public investment programmes and the annual process of budget preparation. We call upon the EAP Task Force to deepen its work on existing finance strategies in the water sector and to broaden the methodology to other areas of environmental policy and to extend its application to other NIS. We call upon the Project Preparation Committee (PPC) to intensify its efforts to support projects identified in national environmental finance strategies as well as other strategically important water projects in the NIS. PPC should prepare a list of projects in the water sector in the NIS and present them to the EAP Task Force.

Reform of the Urban Water Supply and Sanitation Sector

Although investments are urgently needed, they will not, by themselves, be sufficient to reverse the deterioration in urban water services, and the consequences this would have on human health and the environment. Moreover, increasing financial resources in the present institutional and regulatory circumstances may reinforce inefficiencies which currently exist, weaken the urgent need for change and increase the costs of remediation. The financial resources required to address the scale of the existing problem will not be available without significant structural reform in the urban water sector.

Various initiatives have been taken to reform the urban water sector, often with donor and IFI support, but the results achieved have been mixed. It is doubtful whether such initiatives can be replicated so as to address the scale of the problem that exists in most NIS without more fundamental reform. A new policy and institutional framework is urgently needed to facilitate and support investments, and to place the management of water supply and sanitation services on a stronger basis. This will be essential in order to mobilise necessary financial resources. Ultimately, urban water sector reform should be integrated into river-basin management schemes and other government policies and programmes in the light of sustainable development. Some of the key elements of reform include:

- Decentralising responsibility for water service provision from national to local level and strengthening the related capacity of local authorities, in particular locally-elected governments.
- Transforming vodokanals (water utilities) into autonomous, commercially-run institutions under strict supervision by public authorities.
- Promoting a more balanced development of urban waste water treatment relative to water supply, particularly in small and medium-sized towns.
- Engaging the public directly in the reform process and making adequate provision for consumer protection.
- Establishing the sector on a financially sustainable basis, while addressing the needs of poor and vulnerable households.
- Creating incentives to substantially increase efficiency in the use of water by consumers and in the operation of vodokanals.
- Creating conditions for private sector participation within an appropriate regulatory framework.

We endorse the Guiding Principles for Reform of the Urban Water Supply and Sanitation Sector in the NIS as a basis for further work in this area. We urge all stakeholders to actively support the implementation of the Guiding Principles in the countries of the region, and we recognise the importance of ensuring broad public support for the successful implementation of reform programmes. We call upon the EAP Task Force to use the Guiding Principles as a framework for elaborating a focussed programme of work to support the NIS in reforming their urban water and sanitation sectors. The NIS authorities with direct responsibility for the urban water sector should be invited to participate actively in the development and implementation of this programme.

Successful reform of the water sector will largely depend on the commitment of NIS governments to develop and implement economic reforms in their countries. We call upon national and international donors to increase considerably assistance to the NIS countries to support reforms and priority investments. Successful reform will facilitate the provision of donor support.

Private Sector Participation

Private sector participation in the provision and management of water services is increasing in many parts of the world. We recognise that the private sector can provide finance, and introduce more efficient management and technologies to vodokanals. However, we also recognise that all forms of private sector participation require a strong role for government to protect public interests and rights, and to guarantee the level and quality of service provision. The general public and NGOs should be actively involved in this process.

Currently there are many obstacles to private sector participation in water sector in the NIS – policy, legal, institutional, financial and cultural – as is witnessed by the small number of such initiatives. Nevertheless, as reform of the water sector proceeds, private sector involvement may provide an increasingly attractive option for providing water services. We call upon the EAP Task Force to help facilitate dialogue between NIS governments, the private sector, the general public and other stakeholders, to monitor progress, and to disseminate the results of successful initiatives.

Public Participation and Consumer Protection

The public should be actively engaged in the process of reforming the urban water system from the very start to receive timely and exhaustive information, to offer citizens an opportunity to express their views and to participate in the decision-making process. In this way the rights of the citizens to a healthy environment and the consumers' rights to clean and affordable water can be ensured. It is of paramount importance to establish co-operation with NGOs which are the most active and best organised part of the public. NGOs are able to play a greater role in the process of drafting, implementing and monitoring water reform plans at all levels. NGOs can explain to the public the necessity of reforming the water sector, disseminate information on the reforms underway and promote the dialogue between governmental bodies, local authorities, vodokanals (water utilities) and the public.

Next Steps

The worsening, and in some NIS, the already critical condition of urban water services means that the continuation of current trends is not an acceptable option. Failure to take more effective measures will result in a further deterioration of water services and potentially serious impacts on human health and the environment in the NIS. Thus, the first goal must be to stop the deterioration in water and sanitation services as a step toward the sustainable provision of good quality, reliable water services delivered at least cost to the population. We acknowledge the importance of all stakeholders taking the steps required to achieve this goal and supporting the actions we have recommended. Achieving these goals will require strengthened and sustained efforts over many years. NIS governments, donors and IFIs are key in facilitating these processes and should work in active partnership with all stakeholders, including the public and NGOs. New RECs could play a particular role to facilitate multistakeholder debate and capacity-building in this field. We encourage participants to the London Environment and Health Conference, June 1999, to ratify the Protocol on Water and Health, adopted by Ministers at that Conference.

We call for these Joint Conclusions to be presented to the Kiev "Environment for Europe" Ministerial Conference. We urge all stakeholders to work together to prepare concrete proposals for the water sector in the NIS region, in co-operation with the PPC and EAP Task Force, taking account of experience in Central and Eastern Europe. With a view to the Rio +10 and the International Conference on Freshwater (December 2001, Bonn) we will assess progress in urban water supply and waste water management within a broader perspective of local, national, and regional management of freshwater resources and consider a more action-oriented approach. We request the EAP Task Force to monitor progress in water sector reform, including implementation of investment projects, and to submit a report on this subject to the Kiev Conference. We invite Ministers at the Kiev Conference to commission the preparation of an objective assessment of progress in stopping and reversing the deterioration in the urban water services in the NIS. Such an assessment could be discussed at a conference of stakeholders no later than 2005. There is an important need to address rural water issues.

In carrying out the work we have recommended, the EAP Task Force should make every effort to draw upon, and contribute to, related efforts in other international bodies.

Chapter 2

BACKGROUND PAPER ON VALUING ENVIRONMENTAL BENEFITS AND DAMAGES IN THE NIS: OPPORTUNITIES TO INTEGRATE ENVIRONMENTAL CONCERNS INTO POLICY AND INVESTMENT DECISIONS

I. Introduction

Background

At the "Environment for Europe" Conference held in Aarhus in June 1998, environment ministers stressed the critical importance of integrating environmental considerations into sectoral and economy-wide policies to enhance environmental improvement in the region.

Recognising that major decisions affecting the environment in transition economies are being made outside the environment ministries, the EAP Task Force emphasised the need to improve the dialogue between environmental authorities, at all levels of government, and their counterparts in other national ministries, in regional government and at the municipal level who are involved in making decisions on public investments. Awareness of the value of environmental improvements and the costs of degradation appears to be a critical factor for motivating other ministries to consider the environmental impact of proposed restructuring initiatives. That is why the Task Force members approved a special activity designed to enhance local capacity to use environmental valuation in decision-making in the NIS.

The World Bank has taken the lead in implementing this activity, which consists of three phases:

I. Review of recent efforts to apply modern valuation methods for assessment of environmental benefits and costs in the economies in transition;

II. Training of technical experts and decision-makers in applying modern environmental valuation methodologies and assessing their suitability to the NIS context;

III. Pilot studies supported by international experts to demonstrate the potential of environmental valuation, both as a tool in the decision-making process, and as a means to improve skills through on-the-job training.

This paper summarises the findings of Phases I and II and reports on the progress of two pilot studies under way. It illustrates ways to improve the process of prioritising public investments through consideration of the monetary value of the costs of environmental degradation and of the benefits of environmental improvements.

This paper consists of six sections. Section II of the paper discusses the potential advantages of using economic valuation of environmental impacts to prioritise problems and to agree on cost-effective means to resolve them. Section III illustrates recent experience and discusses technical and institutional constraints on application of modern valuation methodologies in the NIS. Section IV describes the potential role of environmental valuation in decisions concerning water supply and water quality. Section V describes the progress of the pilot valuation studies undertaken under the EAP Task Force Program for the NIS. Section VI offers conclusions and recommendations.

The key message

The economic value of health and productivity benefits of environmental investments is often overlooked during appraisal. This delays financing of projects which, by preventing further deterioration of human health and natural resources, could bring about substantial savings to society. However, well-developed, tested methodologies for estimating the economic value of environmental costs and benefits exist and can, with certain limitations, be applied in the NIS.

Evaluating environmental impacts in monetary terms can enhance the ability of environmental authorities at national and at local level to hold meaningful dialogue with economic and finance ministries on the cost of environmental degradation to the national economy and on budget allocation to environmental improvements. Public awareness of the value of environmental benefits and costs can also improve understanding of the trade-offs between environmental and other investments and help in the process of prioritisation.

II. Economic valuation as a tool for integrating environmental concerns in decision-making

Economic Valuation can help decision-makers faced with difficult trade-offs

NIS economies share a number of common characteristics, which make environmental protection a highly controversial issue, ranked relatively low on the political agenda. The environmental disruption that took place in the past is widely acknowledged, and continues to cause severe damage. But people and firms tend to consider environmental damage a secondary problem compared to the other hardships of the transition period: high inflation, persistent unemployment, lessened social cohesion, financial arrears, adverse terms of international trade, and political instability, to name just a few.

Popular conviction that the goals of economic development are in conflict with those of environmental protection effectively undermines attempts to address environmental problems. Frequently cited arguments for inaction include:

- "Everybody grows first and cleans up later" – and this should apply to the NIS as it has elsewhere. What we need now are jobs and growth in order to generate the resources necessary to address environmental problems later.
- We cannot enforce the emission/discharge standards because we would have to close many factories and leave whole cities without means of livelihood.
- Individual incomes are low, so the prices of energy and water should be kept at affordable levels through subsidies.
- Public resources are very limited, and little money is left for addressing environmental problems after more important or urgent programs get financed.

Each of these arguments illustrates the difficult choices concerning the use of natural resources and the priorities for improving environmental quality in the context of the transition to a market economy. Environmental valuation provides a tool to advise decision-makers which interventions are most urgent and which can wait. It helps direct resources to those activities, which achieve environmental benefits or mitigate environmental damage at least cost.

Economic growth and creation of jobs is undoubtedly a first-order priority for the NIS. It is also true that some economies grew fast while heavily disrupting their natural environments. This was partly a result of ignorance of the damaging consequences for human health and for the livelihoods of those who depended on the use of natural resources, and partly a result of the lack of effective environmental regulations. With hindsight, however, it has been estimated that society in these countries ended up paying more in terms of damages and their remediation than it would have had to pay had environmental regulations been in place and implemented at the time of rapid growth (see Box 1).

The NIS economies are at a stage of development where the greatest improvement in environmental performance is expected to result from improved management of scarce resources and technological modernisation linked to new investments. If a proper environmental regulatory framework is in place at

> **Box 1. When should environmental expenditures be made?**
>
> Most industrialised countries have faced the need of large expenditures to remedy the effects of past environmental neglect. Their experiences suggest that it is very expensive to undertake a rapid transition from neglecting environmental goals to giving them high priority. Countries that wish to catch up with the environmental performance of OECD members will find it more effective and less expensive to undertake the transition over a long period – starting sooner, rather than later.
>
> Recent analysis in Japan compared air pollution abatement costs to the cost of health damage cause by air pollution. The comparison took a rather conservative approach, in that the pollutant studied was sulfur dioxide, a substance less damaging to human health and more expensive to control than some other pollutants. Even so, the study concluded that Japan would have been better off had it taken action 5-10 years earlier than it did.
>
> _____
> Source: World Bank (1997) "*Can the Environment Wait? Priorities for East Asia*". Washington DC.

the beginning of the modernisation effort, and if enforcement is credible, then environmental considerations will be incorporated into investment decisions and adverse environmental impacts, possibly entailing huge losses, will be prevented.

Enforcement of environmental regulations is often relaxed out of concern for social problems that might arise, especially in the case of one-company towns. Economic valuation can help determine whether environmental health damages are so large that allowing a polluting business to continue operation would result in increased spending on health care and unacceptable risks to public safety. Sometimes, however, the social cost of unemployment and the associated potential health damages from under-nutrition are larger than the potential environmental damages associated with continued operation, and the enforcement authority needs to consider these effects.[1]

Economic valuation can also help determine the environmental cost related to excessive natural resource use and the willingness of people to pay for environmental services. There have been only a few studies in the NIS on how much people are willing to pay for services such as sanitation, provision of potable water, electricity, etc. Local authorities tend to set prices for such services below their costs, and probably below what households would be willing to pay should they be asked about their preferences. Keeping these tariffs very low increases the demand above its socially justifiable level and puts excessive pressure on natural resources and public funds for the provision of these services.

It is certainly true that financial resources are severely limited and that expenditures for environmental improvements compete with other urgent development needs. Hence the need for prioritisation and for a justification of environmental benefits based on their net benefits or their contribution to the well-being of the population There is a risk of focusing only on restricted financial resources as the root problem and of postponing environmental improvements until a more prosperous future. A clear assessment of the value of environmental damages or benefits can help find solutions that can be implemented now without substantial financial resources, or in some cases can help save resources (see Box 2).

Economic Valuation of Environmental Impacts – Brief Overview

Governments, especially ministries of finance and economy, usually have much clearer perceptions of the costs associated with implementing environmental regulations than of the benefits of these measures or the costs of inaction. One reason for this is that assigning monetary value to these benefits and losses is not easy. Economists recognise that there are several components of value, some of which

> **Box 2. Fighting infectious diseases in Orlovsky District, Rostov Oblast, Russia**
>
> The Orlovsky District Administration was concerned that the frequency of hospitalisations due to gastric-intestinal diseases in the district was far higher than the oblast average and the average in neighbouring areas. Expansion of the infectious disease ward of the district was considered. Increasing the number of beds from 60 to 90 and installing additional equipment would have cost approximately US$2.2 million.
>
> However, analysis of the causes of increased incidence of infectious diseases showed that its main cause was the unsatisfactory quality of water abstracted from ground sources. A chlorinating station for water disinfection was built in 1999 for a cost of US$220,000. In a few months, the frequency of cases of acute gastro-intestinal disorders drastically diminished, and there was no longer a need to expand hospital capacity.
>
> *Source:* World Bank staff, North Caucasus Regional Policy Unit, Russia Environment Management Project.

relate to using things and some to merely knowing that things exist. The former have been named "use values" and the latter "non-use" or "passive use" values (see Box 3).

Environmental valuation involves placing monetary values on environmental goods and services and on changes in environmental quality resulting from certain actions or inaction. Unlike other goods and services, environmental ones are not subject to market transactions and their value is not revealed by market prices. However, economists have developed ways to estimate their value on the basis of physical or behavioural linkages between the indicators of environmental quality and the observed effects on health, productivity, or natural resource assets. **Physical linkage methods**, known also as dose-response methods, rely on a technical relationship between environmental degradation and physical damage, without taking into account the subjective preferences of the affected people. **Behavioural linkage methods** assume that the value of environmental goods and services can be approximated by

> **Box 3. Categories of economic value**
>
> Economic analysts assume that for all practical purposes values are assigned in the process of consumption rather than production. In other words, nothing has a value unless it serves human need either directly or indirectly as a factor of production. The total economic value attributable to the use of the environment and natural resources may be obtained by summing estimates of value for each of the following five major categories:
>
> - **Direct use entailing physical resource extraction.** Examples include consumption of goods provided by biological resources such as timber, fibres, food, medicinal herbs, fossil fuels and other mineral resources.
> - **Direct use not entailing physical resource extraction.** Examples include consumption of services derived from natural resources, such as tourism, recreation, education and scientific research.
> - **Indirect use.** Society benefits from ecological functions that support economic activity and human welfare, *e.g.* waste dissipation/assimilation, climatic functions and water retention provided by forests.
> - **Optional use.** This category reflects known and hypothetical future uses of any type listed above. For example, preserving biodiversity may be viewed as holding an insurance policy against various courses of events and keeping society's options open.
> - **Non-use (passive use) values.** These reflect satisfaction from the existence of a resource such as a mineral deposit, an organism, species or ecosystem, as well as from its bequest to future generations.

people's willingness to pay for improved environmental quality or to escape deterioration. Behavioural linkage methods can be subdivided further depending on whether people's preferences are revealed indirectly, through market behaviour, or directly, through a statement (see Boxes 4, 5, and 6).

A major difficulty in valuing environmental impacts arises because of so-called externalities, *i.e.* those outcomes – both intended and unintended – that affect third parties. They can be either positive (*e.g.*, when a natural resource is conserved or reconstructed for a common good) or negative (*e.g.*, when pollution is emitted). The problem with valuing externalities is that they are not subject to market transactions and, consequently, there are no market prices that can capture the economic values involved. The agents generating these external effects do not bear the full benefits/costs of their actions. Nevertheless these benefits and costs are "real" in the sense that they add/ subtract – perhaps indirectly – from someone's material resources or level of satisfaction.

The most frequently considered environmental externalities are those related to human health. Epidemiological studies have determined the health effects of certain harmful substances discharged into air or water. The quantitative relationships linking pollutant concentrations to morbidity and mortality, called "dose-response functions," are central to carrying out health risk assessments. To establish the health costs of pollution, researchers study patterns of exposure to the pollutant, costs of health care, work days lost due to illness, and, in the cases where the impact is increased mortality, the value of human life.

The costs associated with health externalities are sometimes so significant that costly preventive measures are economically justified. For example, the water-pollution related mortality and morbidity costs in Moldova were estimated at US$60-115 million annually. Provision of piped water of adequate quality from different sources would cost US$23-38 million annually. Thus, investments in improvement of drinking water would be well justified on the basis of potential savings in health costs.

At the same time there are cases, where the perceptions of the costs of environmental damage are unrealistic. For example, a multi-media health risk analysis for Samara Oblast, carried out by Russian analysts, found very low carcinogenic risk associated with drinking water. This finding was contrary to local opinion that leakage from a technogenic oil bed located in the village of Lipyagi posed significant carcinogenic risk. The study results showed that air pollution abatement would be more likely to contribute to lower cancer rates.[2]

Box 4. Valuing the health and safety effects of environmental degradation: physical linkage methods

Techniques that use a dose-response (damage) function relating pollution exposure to health and productivity changes include the **Cost-of-Illness** approach, the **Human-Capital** approach, the **Cost-of-Productivity-Loss** approach, and the Replacement-Cost approach. Estimation of health and productivity effects requires three steps:

1. Risk assessment to determine a relation between indicators of environmental quality and human mortality/ morbidity rates, or productivity changes of bioresources, or material damages.
2. Calculation of physical damage for the specific situation, using the dose-response coefficients estimated in Step 1.
3. Evaluation of the monetary value of calculated damages using prevailing market prices for medical resource costs and statistical estimates of the value of life. A frequent drawback of physical linkage analyses is the lack of scientific knowledge about causal relationships and lack of relevant data to establish one.

> **Box 5. Valuing non-market goods on the basis of stated preference: Contingent Valuation Method (CVM)**
>
> The CVM attempts to measure individual "willingness to pay" (WTP) for environmental improvements by directly questioning a representative sample of individuals. This method has universal applicability in valuing non-market environmental goods, including non-use values such as option and existence values, and has minimal requirements for secondary data. CVM studies are conducted using surveys whose design is important in interpreting the results. Standard survey-design quality-assurance procedures must be followed (such as use of control questions, piloting of the instrument, use of representative samples, interviewer training, etc.). CVM studies have received mixed assessments because of the potential discrepancy between stated behaviour in a hypothetical situation and actual behaviour. However, many researchers feel that conventional biases can be dealt with through better survey design or through careful interpretation and qualification of the final results.

In addition to human health, changes in environmental quality may also lead to changes in productivity or may affect the state of natural resource assets. For example, pollution of water bodies may lead to significant losses in the fishing and tourist industry. A careful consideration of these potential impacts may provide economic justification for measures to mitigate pollution.

Two central concepts in valuing externalities are the individual's "willingness to pay" (WTP) for environmental improvements and the individual's "willingness to accept compensation" (WTAC) for adverse changes in environmental quality. Estimating WTP has important applications in decisions about upgrading and rehabilitating environmental services infrastructure and in pricing the provision of such services.

> **Box 6. Determining the demand for environmental goods and services: revealed preference methods**
>
> **Hedonic Pricing** methods aim to measure the implicit value of environmental quality as revealed by individuals' preferences for related market goods. Commonly used markets are the housing and labour markets. The assumption is that WTP for environmental quality and safety can be inferred from information on price and wage differentials on one hand and environmental risks or characteristics associated with a specific area or job, on the other. The data and econometric knowledge requirements for conducting a hedonic pricing study are quite burdensome, however. Moreover, in heavily regulated or otherwise distorted housing and labour market, the available market data will convey erroneous information.
>
> **Travel Cost** methods are widely used to estimate the amenity and/or recreational value of outdoor recreational sites such as parks and lakes. The underlying assumption is that people's demand for the recreational site is revealed through their willingness to spend money and time travelling to the site. Data on travel costs and other socio-economic characteristics of users is collected through site surveys and aggregated to produce the aggregate demand curve. This method is less applicable to urban amenities requiring only short trips.
>
> The **Averting/Mitigating Behaviour** approach studies the costs (monetary and opportunity) that people incur in order to avoid adverse environmental impacts. Expenditures are usually for substitute goods (e.g., buying bottled water instead of using tap water) or for activities reducing the associated environmental impacts (*e.g.*, the cost of soil-erosion prevention). The underlying premise is that an individual's perception of the cost imposed by adverse environment quality is at least as great as the individual's expenditure on goods/activities to avoid the damage.

While productivity changes can be evaluated with relative ease by considering changes in prices of market goods, estimating the amenity and recreational value of natural resources (water bodies, national parks, species), on the other hand, tends to be a more complex task because it involves subjective values. There are a number of techniques for assessment of subjective values, and the information obtained in such studies can be especially useful in decisions about developments of forestry sector, agriculture, road infrastructure and the like, involving consideration of competing commercial and non-commercial uses.

Table 1 below summarises the broad range of circumstances where assessment of the values of environmental impacts can provide useful information to decision-makers at national, municipal, and project level. It is important to bear in mind that environmental impacts are only part of project/policy outcomes, and therefore environmental valuation should be viewed as only one element in the process of decision-making, complementing engineering, financial, and economic analyses of proposed projects and policies. Ultimately, the decision-making process is political, and it is important that politicians be well informed about the potential consequences, both positive and negative, of their decisions.

Table 1. **Suggested applications of environmental valuation at different levels of decision-making**

	Priority setting	Policy design/Analysis	Investment selection/Appraisal
National/ Regional level	• Federal budget allocation for environmental protection, public health and sustainable development; • National and/or regional priority-setting to allocate scarce financial and human resources; • Assessment of effects of policies related to natural resources (*e.g.*, licensing, trade, and property rights).	• Setting environmental standards/ fees/taxes; • Support for regulatory actions and legislation through estimation of environmental benefits and losses which would be incurred by inaction, late adoption or elimination of laws/regulations.	• Appraisal of large-scale public investment projects which have major impacts on environment and natural resources; • Environmental impact assessments and state environmental reviews.
Municipal level	• Municipal budget/local environmental fund allocation; • Negotiation with the region for money from a regional budget for environmental protection.		• Verification of environmental effects for projects receiving concessionary finance from municipal environmental funds; • Negotiation of intermediate compliance targets for large polluters; • Land use planning and development; • Insurance against environmental risks.

III. Recent practices in using environmental valuation in decision-making in the NIS

The idea that environmental impacts of proposed investments have to be accounted is not entirely new to the NIS. In 1987, the State Committee for Environmental Protection of the former USSR adopted guidelines for calculating the cost of adverse environmental impacts. The so-called "Metodika"[3] has limitations in its ability to capture true value. It was used primarily for assessment of environmental fees to be paid by polluting enterprises. Valuation of environmental impacts was not used for priority setting or regulatory decision-making at the national level. In recent years, however, modern methods have been applied to evaluate environmental impacts (both positive and negative) at the project level and to support regional and national priority setting. This chapter will provide illustrations of recent experience and will discuss the technical and institutional constraints on applying modern methodologies for valuation in the NIS countries.

The NIS "Metodika" for damage estimation: description and limitations

The main idea of the "Metodika" was to aggregate different pollutants into standard units and to assign economic value to a unit of aggregate pollution. The national guidelines (1987) used specified conversion coefficients to convert metric tons of emissions/discharges into "standard" tons of emissions/discharges. These conversion coefficients were based on epidemiological data on the relative harm caused by pollutants, emission source location, height of pipe, speed of emissions and climatic conditions. The system proved to be too complicated for implementation, so it was simplified, and in the process its scientific basis was eliminated.

The most controversial step was the valuation of a "standard" unit of aggregated pollution. For air pollution, the value was RUR 2.6 per "standard" ton, for water pollution about RUR 400 per "standard" ton. There was an initial attempt to base these values on the economic value of human life and human health, damage to agriculture, the forest sector, fishing, or productive assets, and so on. But, for all practical purposes, the values were assessed with a view to raising the necessary resources for mitigating actions. In other words the valuation reflected planned "willingness to pay" for pollution reduction instead of real economic damage from this pollution.

This method tends to underestimate the costs of environmental damages. For example, estimates of the costs of air pollution damage in Sverdlovsk Oblast based on dose-response methodology appear to be an order of magnitude higher than the estimates produced using the "Metodika" (see Box 7). Another major shortcoming of the "Metodika" is the lack of consideration of exposure data, which means that the damage costs associated with a ton of particulate matter emitted by a factory would be the same for a factory situated in the middle of a major city and for a factory situated in less densely populated areas.

Box 7. Comparison of estimates of environmental damage costs

As part of the analytical work to define priority environmental actions in the Regional Environmental Action Plan for Sverdlovsk Oblast, a team of Russian and international specialists attempted to assess the cost of environmental damages from pollution. For air pollution, estimates were generated in two ways: i) using the current Russian methodology of damage estimation, and ii) using the OECD-recommended methodology based on dose-response functions. Below are the results for three major pollutants:

Rate of health damage from pollutant emissions into ambient air

Damage (thousands of dollars/ton)	OECD method	Methodika
Particulates	5	0.2
Lead	123	13
SO_2	0.2	0.1

The team concluded that the Russian methodology consistently undervalued the costs of environmental damages. The discrepancy was probably overstated, because the value of loss of life used in the estimation process ($0.35 mln), based on OECD estimates, was considered high by Russian experts. Even so, the team was unanimous in finding that the Metodika calculations underestimated environmental damages in the oblast, leading to a lower-than-optimal level of investment in pollution abatement.

Source: Ural Regional Environmental Action Plan, produced as part of technical assistance under World Bank financed EMP project in Russia, 1998, pp. 19-22.

Applications of modern methodologies in the NIS

In the past 10 years, valuation studies have been carried out in the NIS (primarily Russia), mostly in the context of capacity building efforts such as development of national and regional strategies or enhancement of technical skills. These studies were usually supported, both financially and intellectually, by international experts and institutions. Annex I describes the methods used and results obtained in studies reviewed for this report. Some studies (*e.g.* the study of Severstal Iron and Steel Works in Cherepovets) have been conducted in co-operation with decision-makers and their results have been incorporated into investment decisions. Other studies have been met with scepticism or open resistance from authorities (*e.g.*, surveys of Moscow drinking water). The main lessons learned from the limited experienced gained so far with valuation studies in the NIS has been the importance of careful interpretation of the results and the need to improve communication between technical experts and decision-makers.

In 1995, the World Bank and the Government of the Republic of Moldova prepared a National Environmental Action Plan that used environmental valuation to set national priorities. Estimates of the costs of health and productivity impacts formed the basis for both institutional and sectoral recommendations. A large share of health costs appeared to be due to unsafe drinking water (see Box 8).

Most NIS studies attempt to evaluate the health costs of air pollution in major urban centres. The dose-response functions linking air pollutant concentrations to mortality and morbidity are well developed in OECD countries, as are standard procedures for adjustment of these functions to different contexts. Since people can take only limited measures to avoid polluted air (*i.e.*, people who live and/or work in a particular area must breathe the air there), exposure can be determined with reasonable certainty and health costs can be estimated relatively quickly.

Studies have consistently found air pollution to be a serious health risk. In the case of Sverdlovsk Oblast, Russia, annual costs were estimated at up to US$1.1 billion, about 13% of the oblast GDP in 1996.[4] Another study, conducted by HIID in Volgograd, found that net benefits in terms of health improvements proposed particulate control measures would be about US$40 million. Indeed, the cost per life saved of the measures suggested for Volgograd would be only US$90-200.[5]

A comprehensive air pollution health risk assessment was carried out for the Severstal Iron and Steel Works in Cherepovets (Vologda Oblast, Russia) by Russian experts working with public health experts from Harvard University. The results of the study were incorporated into cost-effectiveness analyses for 19 investment projects proposed under the Federal Program for Environmental and Population Health Remediation in Cherepovets (see Box 9).

Box 8. **Using valuation for environmental priority setting at the national level – the Moldova NEAP**

The health impacts of water and air pollution were quantified. Polluted drinking water emerged as the major environmental health problem in Moldova. Estimated average annual costs resulting from water pollution were US$60-115 million due to premature deaths and US$6-12 million due to illness. The health benefits of supplying piped water of adequate quality from alternative sources were approximately 3 times larger than the average investment costs.

Premature death and illness associated with exposure to air pollution (TSP, SO_2, Pb) were estimated to cost the economy US$18-33 million, of which US$17-30 million was attributable to particulates and US$1.5-3 million attributable to lead. Productivity losses of agricultural land erosion were estimated to cost US$45-55 million annually.

Source: National Environmental Action Plan, World Bank and Government of Moldova (1995).

Box 9. **Using valuation for prioritising investment at the plant level – Severstal Iron and Steel Works**

The study was part of a technical assistance program financed by a World Bank loan to the Government of Russia under the Environmental Management Project. The objective was to advise the plant and municipal authorities on optimal investments under the Federal Program for Environment and Population Health Remediation. A joint team of Russian and international consultants assessed the potential reduction of health risks to the population of the city of Cherepovets associated with each of 18 proposed groups of investments. Using this data, together with data on the cost of each investment group, it was possible to rank the investment groups in terms of their cost-effectiveness in reducing mortality, respiratory and cardio-vascular disease.

The graph below summarises the findings for respiratory disease. Investment group 17 is the most effective in reducing the number of cases, but is not necessarily the most cost-effective. The team's recommendation to Severstal management and the Cherepovets authorities was that a combination of investment groups 2, 3 and 8 would achieve significant health benefits at lowest cost per life saved/disease avoided.

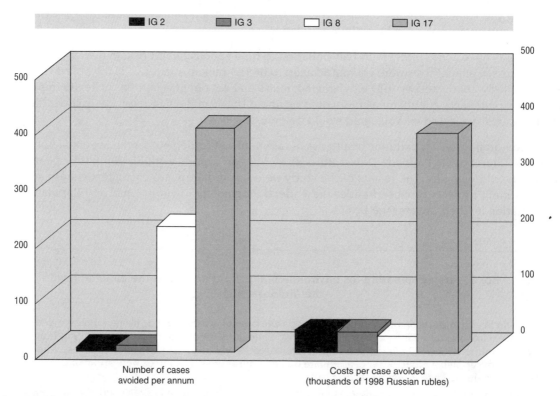

Investment Groups:
IG2: Improve operation of coke dry-quenching unit #2 to eliminate release of coke gas.
IG3: Improve operation of coke dry-quenching unit #3 to eliminate release of coke gas.
IG8: Improve refractory lining and provide gas/waterproof hatches for coke oven batteries.
IG17: Aggregate gas by-products recovery and treatment in selected units.
Source: World Bank Russia Environment Management Project, *"Economic analysis of industrial pollution abatement investments in Severstal Iron and Steel Works in Cherepovets"*, Executive Summary. Moscow, 1999.

Constraints to a wider use of economic valuation of environmental impacts in the NIS

The constraints on a wider application of environmental valuation in the NIS are both technical and institutional. Technical constraints, *e.g.*, data availability, data quality, price distortions, etc., impose limitations on the results and influence how the results should be interpreted. They can in many cases be overcome by careful design and execution of the study. Institutional constraints, on the other hand, make implementation of the valuation studies more difficult or limit the use of the results in the decision-making process. Some of the constraints discussed below are not exclusive to the NIS, so a sizeable body of research already exists on how to deal with them. However, some apply to an NIS setting to a critical degree, and may impair the quality and accuracy of results more seriously than they would in developed market economies.

Information gaps present an important challenge since many of the cause-effect links between environmental problems and outcomes (e.g. the link between water or air pollution and health effects) are only partially understood or measured. If data on levels of air and water pollution, numbers of people exposed, and incidence of disease or premature death are not available, it is difficult to apply the traditional dose-response approach to valuing changes in health outcomes due to changes in pollution levels. The required data can be collected, but this is a time-consuming and often expensive endeavour. There are ways to transfer estimates of dose-response relationships from other locations to produce rough but quick estimates of pollutant impacts – these are useful only if relatively large margins of error are acceptable.

In the NIS, both the existence and the reliability of data present problems. The data necessary for health risk assessment is fragmented, because agencies responsible for monitoring different indicators are under the jurisdiction of environment, public health, or public utility authorities and incentives to link databases are low. Some key pollutants are not monitored or, if they are, the measurement techniques are outmoded. For example:

- Information on total suspended particulates (TSP) is of limited use because it is not known what fraction of particles are smaller than 10 or 2.5 microns (PM_{10} or $PM_{2.5}$), the particle size that causes most health damage. There are no monitoring data of 24-hour or 1-hour concentrations of sulfur dioxide, so analysing acute effects of exposure to high concentrations of sulfur dioxide is not possible, even though brief exposures to high concentrations are more likely to cause health damage than long-term exposures to low "annual average" concentrations.
- There is no systematic monitoring of water at the tap. Yet, much contamination occurs in the distribution network after water leaves the treatment plant, and without data on tap water quality there is little way of estimating actual exposures.

A second information constraint is related to the low *public awareness* of the types of natural resources at risk from degradation, and of the potential negative impacts of air and water pollution on health and natural resources. This is important, because many valuation measures are based on willingness-to pay (WTP) surveys (not necessarily just on direct measures of health or productivity outcomes) and such measures are more accurate when the population is knowledgeable about the situation. Public information is very important in this case, especially understanding of the cause-effect links, and of the nature of threats.

Linked to information and knowledge is another major factor: the *income levels* of the population. Willingness-to-pay is clearly linked to ability-to-pay, and this in turn is largely determined by income levels. Poorer people have fewer options and are naturally less willing to sacrifice some immediate income in exchange for longer-term benefits to health or sustainable ecosystems. It is important to recognise that WTP is captured at a specific moment of time and may change with circumstances. As incomes grow, both demand and willingness-to-pay for environmental improvements will grow. Hence it is critical to prevent irreversible changes, such as species extinction, when incomes are low.

On the institutional side, the major constraints include the lack of capacity and resources to carry out the studies, lack of transparency in investment decisions, and decision-makers' perceptions that the uncertainties involved in the estimation process render the results too arbitrary to serve as grounds for decision-making. These constraints can be overcome by adopting a set of standard guidelines on the use of environmental valuation in net benefit analysis and by developing local capacity for undertaking valuation studies. Some preliminary guidelines were developed at the training workshop in Moscow and a small task force was formed to elaborate them and present them in a form appropriate for adoption (see Box 10).

> **Box 10. Guidelines for economic valuation studies:**
>
> 1. *Start the analysis simply, with the most easily valued environmental impacts.* Any analysis will quickly become complicated, so it is important to focus on the major environmental issues, and especially those where valuation is most feasible. This often means a focus on production or health impacts. In other situations it may be loss of recreational benefits.
> 2. *Recognise the symmetry between benefits and costs.* In the case of air or water pollution, health costs avoided (including costs of treatment, drugs, and lost work-days) are an important measure of the benefits from pollution reduction.
> 3. *Always carry out the analysis in a with-intervention and without-intervention framework.* The correct comparison is not between now and some time in the future, but rather between what would be the situation with the intervention, versus what would be the situation without the intervention. In some cases conditions may worsen over time, even with the proposed intervention, but would have become even worse had no action been taken.
> 4. *State all assumptions explicitly, and identify data used and any additional data needs.* Others can assess the analysis and results only if they clearly understand the assumptions and data used. Such understanding will allow replication of the analysis using alternative assumptions and/or data.
> 5. *Finally, valuation studies should be well documented and should pass peer review.* This step helps ensure that the results are credible and can be used for policy analysis.

IV. Applications of environmental valuation in decisions concerning water pollution control and urban water supply

A number of complex decisions at the national and local levels influence the quality of water resources and the safety of drinking water delivered to residents. Valuation of environmental impacts can be used to assess the efficiency of proposed or enacted regulations, to calculate the net present value of proposed investments in water infrastructure, or to estimate the willingness to pay for municipal water services. This section will relate experience from the United States, United Kingdom, and Eastern Europe to illustrate the potential role of environmental valuation in regulatory and investment decisions related to water supply. It will also discuss the limited NIS experience in estimating willingness to pay for water quality improvements.

Benefit cost analysis of drinking water quality and effluent discharge standards in the US

Beginning with Ronald Reagan, all U.S. presidents have issued Executive Orders that require Regulatory Impact Analyses (RIAs) of regulations costing US$100 million or more annually.[6] As a result, benefit-cost analyses have been conducted for all major environmental regulations issued since 1981. The *use* of these benefit-cost analyses in setting environmental standards is limited by the laws under which the United States Environmental Protection Agency (USEPA) operates.[7] Nonetheless, the benefit-cost analyses that have been conducted show how regulations *could have been improved* had the USEPA been able to use the benefit-cost analyses in setting standards.

Benefit-Cost Analyses of Drinking Water Standards Under the Safe Drinking Water Act. Until passage of the 1996 Safe Drinking Water Act, USEPA was to set maximum contaminant levels as close to health-based goals as feasible. In determining whether health-based goals were feasible, the USEPA could consider compliance costs, but formal benefit-cost analysis was not allowed. The 1996 Safe Drinking Water Act changed this guidance, allowing USEPA to compare the benefits of a proposed standard with its costs. A Regulatory Impact Analysis, was conducted for lead in drinking water. The analysis was useful in helping the agency choose among alternate standards, and demonstrates the type of information that a good benefit-cost analysis can provide (see Box 11).

> Box 11. **Regulatory impact analysis of lead in drinking water**
>
> Levels of lead in household drink water depend on *a)* concentrations of lead in water as it leaves the treatment plant, *b)* the extent of lead pipes, fittings or solder in the water distribution system, and *c)* the corrosivity of water, which affects the amount of lead that leaches out of pipes and solder. Setting an MCL for lead, which applies only to water as it leaves the treatment plant, cannot guarantee a maximum level of contamination in the home. USEPA therefore directed water companies to sample water in private homes and to reduce the corrosivity of water if lead concentrations in home samples exceeded a certain threshold. The RIA quantified the costs and benefits of alternate threshold concentrations.
>
> USEPA's 1991 RIA for lead in drinking water focused on the health benefits of reducing lead. These included reduced hypertension, coronary heart disease and premature death in prime-aged men, and fewer instances of lead poisoning and IQ loss in children. Dose-response functions were used to quantify changes in these health endpoints. Heart attacks and premature deaths avoided were valued at US$1million and US$2.5 million (1988 US$) per case. Avoided medical and compensatory education costs were used to value reduced lead poisonings in children, and changes in earnings were used to value IQ gains from reduced lead exposure. Other benefits that were quantified but not monetized included reduced damage to pipes and plumbing from reducing water corrosivity.
>
> EPA evaluated benefit-cost ratios for alternative threshold levels of lead in household drinking water and chose an alternative which would almost certainly yield a benefit-cost ratio greater than 10. By showing positive net benefits from the rule, helped gain political support for it.
>
> ---
> *Source*: US EPA. Regulatory Impact Analysis of Lead in Drinking Water. Washington DC, 1991.

The fact that US law has discouraged use of benefit-cost analyses as a basis for setting drinking water standards should not deter other countries from using such studies to improve the efficiency of environmental regulation. For the NIS, regulatory impact analysis can be helpful in setting interim target standards and in determining whether the compliance costs associated with a particular standard are commensurate with the benefits that it is intended to achieve.

Regulatory Impact Assessment of proposed water quality regulations in the United Kingdom

The UK government was one of the first in Europe to give special consideration to the cost-effectiveness of pollution abatement technologies. The Environmental Protection Act of 1991 requires regulators to consider not just the technical feasibility of mandated technologies but also their cost-effectiveness in achieving water quality targets. Specifically, the Act modifies the Best Available Techniques (BAT) principle to Best Available Techniques Not Entailing Excessive Costs (BATNEEC), permitting regulators to relax the BAT requirement if the additional reduction of effluent associated with BAT is not substantial enough to justify very high compliance costs.

EU environmental directives set water quality standards that member countries are obliged to achieve by a specified date. Member governments have discretion on the specific implementation arrangements. Recently, the UK government proposed a set of new regulations necessary to ensure compliance with the EU Directive on the Quality of Water Intended for Human Consumption (98/83/EC). The Regulatory Impact Assessment for the set of regulations found that the value of expected benefits would be at least GBP 850 million would outweigh the expected cost of the regulations: GBP 490 million for investments (over the period 2000-2005) and GBP 22 million in annual recurring costs. In addition, the analysis demonstrated that the new regulations would not impose a large burden on household users because the expected improvements in efficiency of the water companies would offset part of the charge increases under the proposed regulations.[8]

Although the UK, as a member of the EU, does not have any realistic option but to implement the Directive as it is European Law, this example shows that Regulatory Impact Assessments are an essential part of the public consultation process and can alert sectoral departments and local governments about specific industries and/or areas that might bear a disproportionate burden of compliance costs. For the larger countries among the NIS, where regional authorities have discretion in implementing centrally set standards, such analyses can help direct the enforcement effort to regulations where the benefits are relatively large in comparison to the compliance costs.

Environmental Benefits from Water Infrastructure Investments

The challenge for NIS is to develop water infrastructure rehabilitation and upgrading strategy that can be implemented under severe financial constraints. Costly investments in building or upgrading wastewater treatment facilities are often proposed for the sake of improving environmental quality. Thus, the value of environmental benefits becomes an essential part of the net present value analysis of proposed projects and could have a significant impact on the final decisions.

In general, investments in facilities with high removal capabilities tend to be expensive and to entail high O&M costs. At the same time, engineering studies undertaken in Central and Eastern Europe have shown that significant improvements in pollutant removal can be achieved by modest upgrades, *e.g.* upgrading primary treatment plants by adding chemical treatment.[9] Additional degrees of pollutant removal, especially nitrogen removal, have higher and rapidly increasing marginal costs. Since health benefits depend on the residual pollutant in the water, most health benefits accrue as a result of primary or primary-and-biological treatment. Thus, the additional health benefits resulting from upgrading to secondary and tertiary treatment might not be large enough to justify the high costs.

However, amenity and recreational values are in some cases significant, and even more importantly, productivity changes which result from the investment need to be factored in (*e.g.*, benefits to fishing, agriculture, tourism). The need for a treatment facility and the cost-effectiveness of alternative pollutant-removal technologies both depend strongly on the specific characteristics of the receiving waters and the use of the water source. When there is a severe threat of eutrophication, as is the case in the Black Sea, the potential ecosystem and productivity losses of inaction may be high, rendering the net present value of costly investments positive (see Box 12).

Having cleaner water sources, however, addresses only one cause of poor drinking water. Another pressing issue, especially in the NIS, is cross-contamination: infiltration of sewage water into drinking water pipes. One possible solution involves infrastructure upgrades; another – alternative water distribution arrangements (bottled water, water galleries, etc.). There has not been a comparative study of the cost-effectiveness of these approaches. Economic valuation of the health impacts could be very helpful in defining realistic investment and/or mitigation strategies for the short and medium term. One of the pilot studies described in the next section of the paper is expected to provide insights into this issue.

Determining Willingness-To-Pay for improvements in municipal water supply

Quantifying risks to human health from exposure to contaminated drinking water is more difficult. In part, this is because there are many different chemical and microbial substances that can contaminate water and damage health, many of which cannot be identified. In part, it is because exposure to contaminants varies greatly, depending on a consumer's position in the distribution network (houses far from the treatment plant are more likely to receive contaminated water than those that are closer), and on individual choices to boil, settle, filter or chemically treat water at home, or to drink bottled water.

CEE researchers and policy makers have often adopted modern valuation methods to estimate the value of environmental improvements. A review of USAID-sponsored valuation studies in CEE[10] reveals that the whole range of available approaches has been applied and, in several cases, multiple approaches have been used to study WTP for improvements in municipal water services.

The Contingent Valuation Method (CVM) seems to be the most widely applied. Part of the reason for this could be CVM's relatively low requirements for secondary data, together with the existence in the region of fairly good and reasonably-priced capacity for survey design and administration. Some

> Box 12. **Benefit cost analysis of a proposed investment program for nutrient reduction in the Danube River and the Black Sea**
>
> The degradation of the Black Sea, caused largely by heavy nutrient inflow from point and non-point land-based sources, has had significant economic consequences for the littoral countries. A region-wide initiative to promote recovery of the Black Sea produced the Danube River and Black Sea Strategic Action Plan. The overall plan calls for a series of urgent investments in nutrient-removal initiatives, including industrial and municipal wastewater treatment, wetlands restoration and promotion of environmentally-friendly agricultural practices (conservation tillage, animal waste systems, etc.). The program is to be implemented over a period of 20 years.
>
> Under a baseline investment scenario (US$40 million), projected nutrient loads for the next ten years show an upward trend which would lead to significant losses in the fisheries and tourism industries along the Danube River and the Black Sea coast and would impose additional health costs on the population. Under the strategic investment program, nutrient loads are projected to decline, preventing those losses and contributing added benefits to the agriculture sector.
>
> The benefits of the strategic investment program were estimated on the basis of averted productivity losses in the fisheries and tourism industries, cost of averted illness, and productivity gains in the agriculture sector. As shown by the graph, the bulk of benefits of nutrient removal would accrue to the tourism and fisheries sectors, while health improvements play a less significant part. On the cost side, an important item is the O&M costs of the wastewater treatment facilities. The net benefit of the strategic program was estimated at US$216-726 million, and the benefit-cost ratio at between 1.23 and 1.76.
>
> The benefit cost study provided guidance to the World Bank and GEF in the design of a strategic partnership program of investments (US$300 million) intended to catalyse domestic and donor finance to achieve the full investment program identified under the Strategic Action Plan.
>
>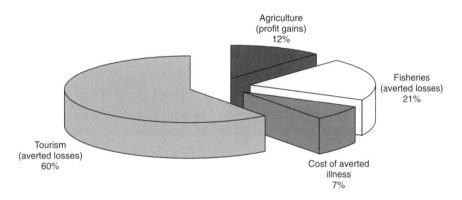
>
> **Distribution of benefits of the Danube/Black Sea investment program**
>
> - Agriculture (profit gains) 12%
> - Fisheries (averted losses) 21%
> - Cost of averted illness 7%
> - Tourism (averted losses) 60%
>
> Source: "Strategic Partnership for Nutrient reduction in the Danube River Basin and the Black Sea: Cost-benefit Analysis," Tijen Arin, draft report, World Bank, ECSSD, 2000.

shortcomings of the CVM method associated with its use in the transition economy context have been identified in CEE applications, but they can be mitigated by careful study design and experienced oversight in the implementation.

Morris *et. al.* (2000) found few instances in which the results of the studies reviewed were introduced into the policymaking process in the CEE. The authors believe that this is due in part to the lack of formal requirements for incorporating environmental cost and benefits into the decision making process, and in part to the authors' failure to design the studies for maximum impact on policy making. The Iasi, Romania study demonstrated how early co-ordination with the policy "players" can enhance use of the study results in policy formulation (see Box 13).

> **Box 13. Contingent valuation of municipal services in Iasi, Romania**
>
> The City of Iasi (400.000 inhabitants) is in north-eastern Romania, not too far from the Moldova border. It is one of the biggest cities in Romania, the site of important heavy industry, power generation, oil extraction and refineries, chemical plants, pulp and paper mills, and timber and wooden product enterprises.
>
> The study examined provision to households of cold water, hot water (including space heat from hot water systems), and municipal solid waste collection. The key valuation method used in the study was the contingent valuation of selected improvements in municipal services as expressed during a personal interview with a household member(s). The WTP estimates were combined with the engineering estimates of the average cost of achieving the service improvements described in the questionnaire.
>
> The study provided a series of recommendations based on the results of the survey. In general, households would be willing to pay more for improved hot water service and improved metering than such service was expected to cost. The WTP for better solid waste collection was roughly equal to its anticipated cost, while the WTP for improved cold water services was less than the anticipated costs. Econometric models revealed that households' willingness to pay for improved cold water service would increase as the economy recovers and incomes pick up. These results were pooled with other observations into a list of recommended changes in price, service levels, and institutional arrangements for municipal services.
>
> The results of the survey provided analytical support for policy makers to adopt new legislation introducing full cost water pricing policy, the beneficiary pays principle, and economic incentives for rational water use. The national water resources management strategy of the Ministry of Water, Forest and Environmental Protection and the new Water Law No.107/1996 took into consideration all the policy recommendations highlighted by the research. Full-cost pricing is an essential condition for the financial autonomy of local water utilities and consequently for the process of local decentralisation.
>
> *Source:* Morris, Glenn *et al.* "Valuation Studies in Central and Eastern Europe: A Stocktaking exercise" Draft April 2000.

In the NIS, a deep-rooted expectation that public services should be provided free of charge as in the past influences the results of willingness-to-pay studies. For example, CVM studies tend to indicate very low WTP for improved drinking water quality. At the same time, parallel estimates based on averting expenditures show a much higher WTP to avoid health risks of drinking water, *e.g.* by buying bottled water. A study, conducted in Danilovo (Yaroslavl Oblast, Russia) estimated WTP for improved tap water as US$0.02 m^3 based on CVM, and US$7 m^3 based on averting expenditure.[11] This can partly be explained by attitude inertia from the past when municipal services were publicly provided at nominal or no charge, and partly by the unreliability of the tap water supply, a factor that would clearly reduce the ability to enjoy the cleaner water. Another study, in Chudovo (Novgorod Oblast, Russia) produced similar results, but the difference was smaller (US$0.08 m^3 using CVM; US$5 m^3 using averting expenditure method), leading analysts to believe that CVM studies produce better results when public awareness of the cost of providing services is higher and generally, in areas where people have greater experience with market transactions.[12]

In sum, CVM studies of WTP for water services are a quick tool with low or no requirement for secondary data which is a major advantage in the NIS. Shortcomings in the method can be mitigated by careful design of the survey instrument, especially by educating the respondents about the situation through the questions (as shown in the Romania case study). Yet, it is advisable to verify results obtained through the CVM by other methods, and to bear in mind that estimates of WTP for public services produced by CVM tend to be lower than WTP estimates obtain by other methods.

V. Pilot studies under EAP Task Force Program

In order to demonstrate the potential use of environmental valuation for decision-making in the context of the NIS, the World Bank, as a partner in the Task Force for implementation of the EAP and with support from the Governments of the Netherlands and Finland, is currently conducting two pilot valuation studies, one in Rostov-on-Don, Russia and the other in Astana, Kazakhstan. The objectives of these pilots are two: *i)* to provide relevant input to decision making and strategy formulation, concerning urban environmental management in the case of Rostov and water resources management in the case of Astana; and *ii)* to build local capacity for environmental economic analysis and provide a relevant learning experience for the NIS-wide initiative on economic valuation of environmental impacts.

The study in Rostov will provide an updated assessment of the health burden caused by air pollution and unsafe drinking water. It will build on and refine estimates made in an earlier study produced for the Greater Rostov Area Strategic Environment Action Plan in 1994. It will also examine possible preventive and mitigating actions, assess their cost-effectiveness and advise local authorities of the findings.

Although at the time of writing of this report the study is still in progress, some interesting results regarding drinking water have emerged. The city planning department has initiated the "Clean Water Program" (see Box 14), which appears to be far more cost-effective in reducing health risks than retrofitting of piped water infrastructure. If confirmed, this result would suggest that the most cost-effective solutions to problems of drinking water quality in the short term may be measures to reduce exposure to contaminants (through the use of defensive measures such as bottled water or clean water from water galleries, filters, boiling and settling), rather than immediate measures to improve water quality in the system.

The Astana pilot study will identify and quantify the potential environmental benefits of alternative interventions for improving the water supply in the new Kazakh capital (see Box 15). As relevant health and epidemiological data are not available to assess health costs by dose-response methods, CVM will

---

Box 14. Clean Water Program in the city of Rostov-on-Don, Russia

The Clean Water Program, implemented under the Mayor's decree is comprised of several measures. The first is construction of a "cold water gallery" in each of the city's eight districts. Five have been completed. The galleries, operating under a public-private partnership, provide 5 liters of water free to everyone who brings his or her own container, and sell quantities in excess this amount for 1 ruble a liter, about ¼ the market price of bottled water. People with incomes below the poverty line pay just 50 kopecks a liter for this water through a program operated by the city's social protection services.

A further measure is installation of filters at schools and hospitals to protect vulnerable people from exposure to contaminated water. The city is also promoting the use of filters in restaurants, cafés, owners of apartment buildings and households. It has established an exhibit of water filters, displaying a wide range of filters intended for different uses and ranging in price from RUR 200 to RUR 200 000. In addition, the program has undertaken a media campaign to raise awareness of the benefits of water filters.

At RUR 300 million, the clean water program presents a quicker and more cost-effective way to supply clean water to the city's residents than investment in piped water infrastructure which would cost approximately RUR 12 billion. These numbers need to be verified, but it is certainly true that such an effort is quite inexpensive and generates immediate results. Additional measures, not directly intended to improve water quality but rather to improve the operational performance and financial sustainability of the water utility, must also be considered, such as installation of water meters and improvement of billing and tariff collection. These will contribute to improved water quality in the long term by reducing excessive demand on the capacity of the water treatment plant and generating resources for investments in upgrading the water distribution system.

> **Box 15. An Assessment of the environmental benefits of alternative water supply interventions – pilot study in Astana, Kazakhstan**
>
> *Background*: Some 1.5 million people live in the Nura-Ishim river basin. Most of them live in Astana, the national capital. This newly created capital city is expected to double its population in the next 10 years, from about 275 000 to about 550 000. The water resources of the Ishim River will not be sufficient to meet Astana's increased demand in the medium term. The Nura River is linked by canal to the Ishim at Astana, but is not currently used due to the mercury pollution from the carbide factory in Temirtau (now closed down). The Government's options are: *a*) bring additional water by constructing an extension to the Irtysh-Karaganda canal or *b*) clean up the mercury from the Nura river. A further consideration is that the Nura eventually discharges into the Kuragaldzino wetlands, an internationally recognised nature reserve.
>
> *Objective of the study*: The principal objective for the policy interventions analysed is to provide a steady and safe supply of drinking water of sufficient quantity and quality to the growing population of Astana. The pilot study will focus on three potential interventions, considering both the demand and the supply side:
>
> - Clean up the mercury in the Nura River. This would provide an abundant reliable source of water to the capital city of Astana, would provide more water for irrigation (150 000 subsistence farmers live along the river), would yield benefits to an internationally recognised water reservoir with rare birds, and might increase revenue from fishing.
> - Bring in water by extension of the Irtysh-Karaganda canal. This alternative would involve rehabilitation of the entire canal (entailing high O&M costs) but could result in benefits for irrigated agriculture along the canal.
> - Undertake policy and institutional reform encouraging water conservation by modifying tariffs and closing leaks, etc. The argument for the need for improved incentives and enforcement could be reinforced by the monetary value of benefits.
>
> The general framework of the cost-benefit analysis is outlined below:
>
>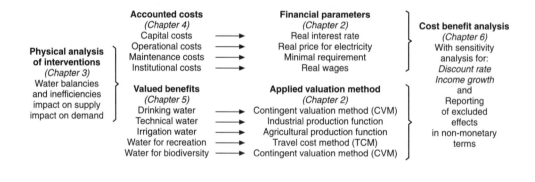

be used. The study will be one of the first attempts to estimate amenity and recreational values as well as non-use values of biodiversity preservation. The results would inform decisions on the long-term water resources strategy for the city.

Though the impact of these two pilot studies on decision-making is yet to be seen, an important implementation lesson has already been clearly demonstrated: support for such efforts on the part of the decision-makers and constant communication between technical experts and regional development institutions are essential. Involvement of decision-makers contributes to better study design through clear specification of the problems at hand, and gives the study a practical, decision-oriented focus as opposed to an academic one. Other lessons have been suggested as well: The quality of the study will likely be enhanced by improved access to data from various institutions; and, finally, the understanding of the issues that decision-makers will develop as a result of interactions with study team will help them more effectively use study results in their decision-making.

Participants in the training workshop in Moscow developed concepts for valuation studies addressing problems relevant to their countries and/or pending decisions at hand. Annex 2 describes some of the proposed pilots, presenting further opportunities for application of modern valuation methods in the context of the NIS.

VI. Conclusions and recommendations

Conclusions

NIS governments face difficult choices in allocating their limited resources to improve welfare, attain economic growth and reduce poverty. Environmental conditions can greatly influence the welfare of people, in particular their health and the sustainability of economic growth. However, the value of environment for society is often underestimated, because it is not easy to express in monetary terms. By contrast, the cost of investments to improve environmental quality is more easily measured and dominates the considerations of decision-makers. As a result, fewer resources are being allocated to improve environmental quality than would be optimal, and the cost of inaction is being ignored.

Economic valuation can help decision-makers to better understand the value of environmental improvements and of the damages that result from inaction. For that reason, economic valuation has become a well-developed field of economic analysis, and although there are special issues in the NIS transition economies, the basic approaches are robust.

Valuation of environmental impacts is, however, **only one factor in decision making**: the results of environmental valuation studies are used to *advise* decision-makers and not to *determine* the decision. In many cases, the chosen option may not be the one with the highest benefit-cost ratio but the one that is more acceptable for political and social reasons. Nevertheless, the economic analysis will show *the cost* of choosing a second-best solution and will enable the decision-maker to assess whether or not this cost is acceptable. Not surprisingly, where the analysis has been developed in close interaction with the decision-makers, the impact on the decisions has generally been greater.

A broader application of economic valuation in the NIS faces some technical and institutional constraints. The technical constraints, such as limited availability and quality of data, or price distortions, can be overcome by careful design and execution of the analysis. Institutional constraints present a bigger challenge because of:

- **Limited regional capacity and expertise** to undertake economic valuations. Despite training efforts by local and international institutions, the circle of regional experts remains small and heavy reliance on costly international expertise continues.

- **Outdated regulations.** Use of more-reliable economic valuation studies in formal decision making is constrained by regulations which still prescribe the use of an outdated "Metodika".

- **Lack of transparency.** Decision-making on investments and expenditures is not subject to public consultation.

- **Skepticism.** Decision makers question the reliability of the results.

Recommendations

A broader application of economic valuation of the environment would benefit not only environmental authorities, but also decision-makers in ministries of finance and economy or at the sub-national level. The private sector would also benefit from information directing scarce resources to areas of major environmental impact.

A broader application will require *i)* a regional capacity-building program, *ii)* revision of procedures for economic valuation of environmental costs and benefits; and *iii)* incorporation of full cost-benefit considerations into the decision making process.

Capacity-Building Program. Building on initial efforts by local and international institutions, NIS countries could pursue a three-pronged strategy:
- Expand the training efforts initiated under the EAP Task Force and funded by donors and international organisations through workshops, training-of-trainers, and partnerships between academic organisations.
- Build a strong, professional network of environmental economists in the region, supported by a common web site which would allow for continuous exchange of best practices and results.
- Increase the number of pilot studies, both for the sake of hands-on training and also to generate a body of robust analytical results which could serve as benchmarks for other NIS studies.

Revisions of procedures for economic valuation of environmental costs and benefits. This would involve the following steps:
- Ministers of finance and environment would establish joint task forces to recommend regulatory changes required to move from an outdated "Metodika" toward the application of modern economic valuation methods in formal decision making. Governments may want to build on the work of a small, region-wide task force that was formed for this purpose in April 2000 at the initiative of several NIS countries.
- The task force would develop a set of guidelines for adoption by the respective governments. It will be important to bear in mind that there is no single, uniform ("cookbook") approach to economic valuation of the environment. While it is important that the method selected be feasible for use in valuing a specific type of environmental good or service, the merit of a valuation method rests more critically on careful design and execution than on the method itself.

Incorporation of full cost benefit considerations into the decision making process. This could be achieved through the following measures:
- Recommend that Economic Impact Analyses be produced and included for the consideration of decision-makers when major projects and regulatory changes are proposed. These studies would include estimates of changes in environmental services affected by the measure.
- Recommend that project and regulatory decisions include an estimate of net benefits for each alternative considered by the decision-makers, as well as an explanation of the grounds for the final choice.

Notes

1. A technical example of such a decision is described in "Employment and environmental protection: the tradeoffs in an economy in transition", Anil Markandya, NIS-EEP Project, Environment Discussion Paper No. 26, HIID, August 1997.
2. Application of risk assessment in Russia (Samara Oblast). International Institute for Health Risk Assessment. Moscow, 1999.
3. USSR State Planning Committee, USSR State Committee for Science and Technology, USSR State Price Setting Committee and the USSR Academy of Sciences "General Rules for Damage Calculation and Cost-benefit Analysis of Abatement Measures". Moscow, 1986.
4. Centre for Preparation and Implementation of International Projects on Technical Assistance: Urals Regional Environmental Action Plan: Executive Summary. Ekaterinburg, 1998; p. 19.
5. Larson, Bruce A. et al. "The Economics of Air Pollution Health Risks in Russia: A Case Study of Volgograd", Environment Discussion Paper No. 38, NIS-EEP Project, HIID, January 1998.
6. Reagan's Executive Order 12291 may be found in 46 *Federal Register* 13193.
7. The National Ambient Air Quality Standards, for example, must be set to protect human health without regard to cost. The statutes that allow benefits and costs to be weighed in setting environmental standards include the Toxic Substances Control Act (TSCA), which regulates toxic substances (such as asbestos) in consumption and production, the Federal Insecticide, Fungicide and Rodenticide Act (FIFRA), which regulates pesticides, and the 1996 Safe Drinking Water Act.
8. UK Department of Environment, Transport and the Regions (2000). The Water Supply (Water Quality) (England) Regulations 2000. Consultations on Regulations. Crown. Available on the Internet at *http://www.environment.detr.gov.uk/ras/index/htm*
9. Somlyody, Laszlo and Peter Shanahan, "Municipal Wastewater treatment in Central and Eastern Europe: Present Situation and Cost-Effective Development Strategies". The World Bank, Washington DC, 1998.
10. Morris, Glenn et al, "Valuation Studies in Central and Eastern Europe: A Stocktaking exercise" Draft April 2000.
11. Cadastr NGO, Russia (1996). "Environmental Valuation of Natural Resources in Yaroslavl Oblast."
12. Centre of Environmental Economics and Natural Resources In Russia (1997). "Contingent Valuation of Drinking Water in Chudovo, Novgorod Oblast."

Water Management and Investment in the New Independent States

Appendix 1. **SUMMARY OF ENVIRONMENTAL VALUATION STUDIES CONDUCTED IN NIS**

Name of the study	Study objective	Method	Results	Policy implications
Health risk analysis/management in Volgograd. HIID, Harvard School of Public Health, 1998.	• To provide guidance on options for least-cost health risk reduction from air pollution (stationary sources) in the industrial city. • Capacity building.	Health risk analysis, involving the following steps: • hazard identification; • exposure assessment; • dose-response assessment; • risk characterization; • risk management; • environmental benefits estimation based on benefit transfer approach.	The five most cost-effective projects, implemented simultaneously, would lead to avoiding 277-966 deaths per year with a total net benefit of US$39.5 million. The costs of the cheapest options were estimated at $90-200 per life saved.	The study was presented to Russian authorities and experts at a series of round tables. The results provided guidelines for environmental funds.
Assessment of the relative cost-effectiveness of options for reducing environmental health impacts of industrial emissions. Centre for Preparation and Implementation of International Projects on Technical Assistance (CPPI), 1998-1999.	• To demonstrate risk assessment and cost-effectiveness analysis as techniques for priority setting at a local scale. • To provide practical advice regarding several environmental investment projects proposed under the "Federal Program for Environmental and Population Health Remediation in Cherepovets".	Health risk analysis. Data on pollution emissions from the plant were used as an input to an air dispersion model (USEPA model was used) to estimate the concentrations of pollutants in the air in the city before and after investment. Morbidity was taken into account as well. Risk reduction investment options were ranked based on cost-effectiveness ratio.	The study indicated control measures would result in potential annual reductions of 300 cases of cancer, 55 deaths and 1 000 cases of acute respiratory symptoms due to TSP and SO_2. Highest priority identified was investments that improved production and for which environmental benefits were an added benefit.	Investments under the Federal Programs for Severstal Iron and Steel Works were undertaken on the basis of study recommendations.
Multi-media health risk analysis in Verkhnaya Pyshma Centre for Preparation and Implementation of International Projects on Technical Assistance (CPPI), in 1998-1999.	• To adjust US EPA health risk assessment methodology to the Russian context. • Capacity building.	Health risk analysis based on US EPA approach. Hazard identification was based on actual monitoring data. Exposure assessment was based on concentrations of pollutants from monitoring data on pollution of environmental media (ambient air, soil, drinking water, and food).	Ranking of environmental risks based on damages to population health: Rank 1 – Lead; Rank 2 – TSP, NO_2, SO_2 (in total); Rank 3 – As, Cd; Rank 4 – Cu; Rank 5 – Other environmental pollutants.	Russian Federal and Sverdlovsk Regional Authorities officially approved this approach. The study produced some recommendations for the Federal and regional SanEpid on means to improve the monitoring process.
Health risk assessment in Samara Oblast Conducted by a group of Russian experts, 1999.	• To assess multimedia health risk in Samara Oblast.	Health risk assessment, US EPA approach. Analysis was based on existing inventory of air and water emissions. Concentrations in air and water were measured with the help of a Russian air dispersion model. The study used benefit transfer approach to estimate WTP for risk avoidance. However, no estimate of the cost of different policies has been provided.	Quantitative characteristics of risk have been developed for the territories studied, for all media. The major medium of cancer risk was atmospheric air, at the level 10^{-3}. WTP for risk avoidance was estimated at US$30 000 per additional non-cancer death and US$300 000 per additional cancer related death.	The study was conducted in close co-operation with Administration and Environmental Authorities of Samara Oblast, providing the authorities with a wide range of legal, institutional and economic policy measures to reduce health risk from environmental pollution.

Name of the study	Study objective	Method	Results	Policy implications
Environmental valuation in Yaroslavl Oblast: Danilovo case study. Conducted by NGO "Cadastr" under the supervision of HIID.	• To assess WTP for improved water supply.	The study applied the CVM approach to estimate WTP of those who do not have tap water for centralised water delivery into their apartments. The survey was arranged as a direct interview with respondents, based on open-ended questions. Respondents were asked to estimate maximum amount of one-time payment they would be willing to make in order to obtain tap water.	Mean of WTP for water supply is about US$0.1/person/month or about US$0.02/m³ of water. 82% were willing to pay up to US$30/household for tap water installation, 8% willing to pay US$100, 4% willing to pay more than US$200.	Study recommendations: 1. Double water tariffs gradually, simultaneously improving water quality; 2. Subsidise low-income households instead of subsidising the local water supply company; 3. Increase the scope of pre-paid additional water supply services for households with the highest incomes; 4. Establish a water supply market with competing tariffs from different water providers 5. Spring water sales
Environmental valuation of drinking water in town Chudovo (Novgorod Oblast), 1998.	• To assess WTP for improved drinking water quality. • Capacity building.	CVM and averting-costs approaches. The main survey was piloted and used both open-ended and discrete choice format. Questions addressed the quality of water, types of avoidance actions and their costs, and, finally, the WTP for specific ways to improve quality of drinking water.	The respondents were willing to pay an additional US$0.5-1.2/m³ per person/month to improve the water quality. WTA was estimated based on adverse costs at US$8-33/month.	The study proposed policy options to improve the water quality situation in Chudovo but was not well co-ordinated with local authorities.
Avoiding health risk from drinking water in Moscow. Conducted by HIID, October 1996.	• To assess attitudes regarding water quality in the city of Moscow.	Survey of opinions of water quality, cold water supply (water quantity), and risk-avoidance actions.	82-88% of Moscow residents were generally satisfied with the cold water quality from Volga and Moskva rivers.	The results of the study were used for capacity-building, as a case study to educate those participating in the Chudovo survey. Several articles were published in the mass media, provoking a very negative response from Moscvodocanal.
Natural resources accounting in Russia: practical experience in Yaroslavl Oblast. Conducted by Yaros-lavl "Cadastr", University of Bath, Institute of System Analysis, Moscow, 1999.	• To develop natural resources accounting system for the region of Yaroslavl.	Direct valuation method based on the UN Guidelines for environmental accounts on monetary and non-monetary levels. This method requires accounting of the stock of each natural resource at the end of the year based on the present value approach.	The total value of natural capital in Yaroslavl Oblast (1996) was US$6.3 billion (19 % of man-made capital). 88% of natural resources value comes from water resources. At the present time the net value of the flow of water resources is very small or even negative.	Results were presented to the oblast authorities nd provoked useful discussions about policy changes.

Name of the study	Study objective	Method	Results	Policy implications
Efficiency and sustainability in natural resources sector of Russia. Regional analysis, conducted by the World Bank and the Higher School of Economics – State University, 1999.	To evaluate natural resources of Samara oblast and to answer the questions: • What are the major components of natural capital and what is their potential contribution to the regional wealth? • What are the channels and efficiency of rent distribution and redistribution? • How can sustainable development be promoted in the region in terms of changing the capital flows and institutional structure in natural resources use?	Direct valuation method to estimate depreciation of the natural capital in Samara oblast, capital value of natural resources, green regional product and net investments. Depreciation of natural capital was estimated based on the change of the capital value of natural capital. Sustainability criteria for the region require non-decreasing Net Regional Product (NRP) or positive Net Investment (NI) as a measure of the potential for regional development.	Land and oil deposits constitute the main natural wealth of Samara Oblast. Their joint capital value is about 7 billion. R in real terms (1994 prices) or US$2 billion NRP started to grow after 1999. Economic depreciation of natural capital makes further growth problematic without substitution of natural capital for man-made capital after 1999. In 2001, economic depreciation of natural capital would be about one third of the depreciation of man-made capital.	
Forest regeneration assessment in Khabarovsky and Krasnoyarsky Krai. HIID and the U.S. Forest Service in collaboration with the Khabarovsk Kray Forestry Administration, Khabarovsk Forest Service and the Institute for Sustainable Communities, 1998-1999.	To select the reforestation methods that are most cost-effective and to identify the sites where they generate the highest returns.	Estimated future growth of different species in different forest zones. Used discounted cash-flow techniques to calculate the cost per cubic meter of regenerated timber.	On average, cost of artificial regeneration is several times higher than costs of assisted natural regeneration. Carbon sequestration investments are viable in most zones of Krasnoyarsky Kray (lower than 10-20 $/ton of carbon). The net revenue of non-timber resources (mushrooms and berries) is approximately $70/hectare.	The Russian Forest Service and regional forest services consider the study one of the first experiences of environmental valuation in the forest sector, which has been financed by public funds without consideration of efficiency.

Appendix 2

SOME PILOT PROJECTS PROPOSED BY THE NIS FOR FURTHER APPLICATION OF MODERN VALUATION METHODS

Valuation of industrial and domestic waste treatment in Sumgait, Azerbaijan Republic

Sumgait (350 000 residents) is located 30km northeast of Baku and has petrochemical industry, metallurgy, power engineering, light industry, etc. The project proposes consideration of different options for wastewater treatment. Currently, the treatment facility is one component of a large chemical plant and carries out only partial waste treatment. Two scenarios are offered for consideration:

- A single treatment system.
- Two separate treatment facilities: one for industrial wastewater and another for domestic wastewater.

Improvement of the water supply in Armenian towns, Armenia

In Armenia, ground water sources account for about 95% of potable water. Water from these sources is of high quality, meeting microbiological and chemical standards. However, an obsolete water distribution system results in substantial water losses (35%-40%) and inadequate supply (on average, tap water is available in households for 1-2 hours per day). The alternatives to improve water supply to be considered are:

- Ensure increased water supply by reducing losses and installing water meters in each household.
- Improve water supply and water quality and introduce a relevant increase in water charges.

The authors proposed to use CVM to evaluate WTP for improved water supply.

Economic benefits of improvement of water supply system and drinking water quality in Kutaisi city, Georgia

Kutaisi (population of 250000) is the second largest city and industrial centre of Georgia, situated in the western part of the country in the Black Sea basin. Although the connection rate to the centralised supply network approaches 90%, presently there are virtually no households provided with a 24-hour drinking water supply. The main pipelines supplying the city with groundwater from external aquifers are 46 km in total length. Project alternatives to be evaluated are as follows:

- No action.
- Rehabilitation of pumping stations.
- Rehabilitation of supply network facilities, including pumping stations.
- Rehabilitation of supply network facilities, including pumping stations, and installation of water meters system to introduce user charges.

The authors proposed to use health risk analysis, other direct valuation methods and CVM to assess WTP.

Assessment of health risk from unsafe drinking water in the Republic of Moldova

Water supply in the cities in Moldova is irregular, and water distribution leakage is at a rate of 30%-35%. There are no household meters. The issue of municipal services privatisation is under consideration. Cost recovery for water treatment and supply is about 30%. The purpose of the project is to elaborate a strategy and establish priorities for reduction of health risks attributable to consumption of polluted drinking water, applying cost-benefit analysis, estimating WTP with a contingent valuation approach, and considering cost-effectiveness.

Project alternatives are:

- Health risk assessment in the current situation (no-action).
- Improvement of water treatment.
- Modernisation of the water distribution system.

Valuation of projects to improve potable water quality in a district of Kyiv, Ukraine

Kyiv has a population of 2.6 million, of whom 100% have access to the central water-distribution system. Water intake is primarily from the Desna and Dnieper Rivers, both classified by the State Committee for Statistics of Ukraine as sustaining an anthropogenic environmental stress. The goal of the pilot project is to assess proposed improvements of potable water quality in a Kyiv district. Project alternatives are:

1. Technical: Measures aimed at improvement of potable water quality:
   - Building a new purification plant at the municipal water supply facility.
   - Renovation of the municipal water supply infrastructure.
2. Institutional: Development of recommendations on improvement of potable water quality, in close co-operation with the municipal authorities, environmental agencies and water management bodies.

The authors proposed application of the "human capital" approach to evaluate the costs and benefits associated with a reduction of health risk due to water pollution and to apply cost-effectiveness analysis to evaluate available options for improvement of potable water quality.

Evaluation of the socio-economic costs and benefits of the development of Kolkheti National Park, Georgia

The Kolkheti Lowland region is located in the Western part of Georgia and is bordered by the Black Sea. The most important parts of the region are its wetlands, which include bogs, marshes, swamps and salt marshes. The flora of the site is diverse, and includes a variety of endemic and relict species. Due to the region's high environmental importance, in order to save the area from further degradation the Central Kolkheti wetlands were recognised as a RAMSAR convention site in 1996. The Kolkheti National Park is intended to become one of the major recreational sites of the Georgian Black Sea coast. The purpose of the project is to evaluate socio-economic benefits related to the establishment and development of the Kolkheti National Park as well as the opportunity cost of foregoing the use of the wetlands and adjacent areas for conventional economic purposes such as pit extraction, hunting, fishing, grazing, infrastructure development.

Valuation of health risk to the population of Bishkek from air pollution related to automobile emissions, Kyrgyzstan

Bishkek, the capital of Kyrgyzstan, has a population of 619 900, more than 13% of the nation's total. In Bishkek, the major sources of pollution are automobiles. About 40% of privately owned automobiles do not meet national norms limiting smoke and toxicity of exhausts. The number of gas stations has increased sharply over the past few years. Automobiles are currently the major source of certain pollutants, including nitric oxides (2.5 MAC), formaldehyde (5 to 10 MAC), benzapyrene (20 to 60 MAC), and suspended particles (3 to 10 MAC). In terms of the gross pollution load, however, the main pollutant is suspended particles (about 50% of total pollution).

The project objective is to assess the economic benefits of improved air pollution control in different districts of Bishkek, based on health risk analysis and estimation of economic damage from transportation.

Project alternatives are:
- No action.
- Regulation of the import of low-quality automobile fuel and automobiles in poor condition, based on testing for PM10, PM2.5, etc.
- Optimisation of the transport network (designing overpasses, erection of noise barriers), increasing the amount of green planting by way of "Ashara" method (sponsored collective work).

An economic analysis would form the basis of recommendations on economic priority-setting for decision-makers.

Chapter 3

BACKGROUND PAPER OF FINANCING STRATEGIES FOR THE URBAN WATER SECTOR IN THE NIS

1. Introduction and key policy implications

This paper summarises the results and draws conclusions from a series of empirical studies conducted in the NIS. Extensive analyses for financing strategies were conducted in Georgia, Moldova and the Novgorod Oblast in the Russian Federation; further studies are underway in Kazakhstan, Ukraine and the Pskov Oblast in Russia. Country studies were implemented by teams of experts consisting of local specialists and the consultants from COWI Consulting Engineers and Planners. Expert work has been monitored by Advisory Groups composed of senior officials and policy makers from the countries concerned. The role of these groups was also to discuss and approve all assumptions and targets used in modelling as well as discuss policy implication of the out-coming results, thereby ensuring full "ownership" of the strategies by the countries. Overall substantive supervision and guidance and the methodological support was provided by the Secretariat of the EAP Task Force. Danish government supported all these activities financially.

Preliminary results have confirmed, in quantitative terms, a serious crisis situation in the water and sanitation infrastructure in the NIS, but also feasible, short-to-medium term actions to achieve realistic and affordable targets. A fundamental challenge to any effort to overcome the water sector crisis in the NIS is to find practical solutions under conditions of very scarce funds in public budgets and extremely limited capacity for additional borrowing. These conditions are likely to last for a considerable period of time in most NIS.

The salient features of the current situation are:

- Current revenues from users are based on so-called "full-cost recovery" principles. However, most water utilities in the NIS *barely cover their operating expenses because full cost is defined in ways that deviate from international accounting standards*. Most utilities are effectively bankrupt by international standards and have no borrowing capacity.

- Capital repairs expenditure depend on public budgets and are generally much lower than what is required to maintain the current value (or average age) of the water and wastewater system. Thus systems deteriorate – often even in the capital cities.

- Policy makers continue to set ambitious targets for the level and quality of the services of the sector even in situations where current service levels are rapidly deteriorating which inflates estimated compliance costs.

- A virtual leak of prioritisation leads to fragmentation of resources and ineffective investment strategies.

- There is almost no private finance in the sector.

Continuation of the current combination of ambitious targets for the level and quality of water services and financing arrangements which involve very low user charges and limited access to funds for capital repairs, is not sustainable. This combination will result in a further unplanned deterioration of the level and quality of services. It is likely to entail *e.g.* low pressures and interrupted supply of drinking water, health hazards because of recurrent incidences of contamination of drinking water and discharge of untreated wastewater even from the largest cities. The main, albeit preliminary, finding of the three country studies is that even to operate and maintain what still works now, will require all three countries to implement demanding packages of policy and institutional reforms. While establishment of more ambitious targets of partial rehabilitation of infrastructure is possible (in addition to operation and maintenance of what works now), the realistic achievement of these targets would require the countries to implement even more demanding packages of policy and institutional reforms to ensure additional finance.

In all countries studied the policy package required to avoid further unplanned deterioration of service level and quality in the urban water sector consists of a few essential, elements:

- There is no feasible, alternative source of finance for operation and regular maintenance expenditure (calculated using international accounting standards) other than user charges. In some of the countries studied (Russia and Georgia) there is scope to raise water user charges within the generally accepted "norms" of what households can afford (*i.e.* 4% of their income). However, increases in user charges should be part of a strategy for improved service provision which has been developed through a participatory process and which makes appropriate provision for poor and vulnerable households. However, in some of the countries studied, user charges are already close to the 4% "norm" of affordability (*e.g.* Moldova and probably Kazakhstan).

- National and local budgets will have an essential role in the short and medium term in financing rehabilitation and capital investments, providing social protection and facilitating access to credit.

- Scarce public funds and donor grants should be concentrated on fewer projects; currently resources are spread too thinly and do not constitute the "critical mass" of rehabilitation investment which is needed to stop the deterioration of water networks.

- IFI projects have an important demonstration and catalytic function, but their number and size will be limited by countries' borrowing capacity. In some cases donor grants and IFI projects should be designed so that operational expenses are financed by public budgets rather than user charges. This underlines the need to concentrate funds on projects that are critical, for example, from a health perspective.

- Private sector investment will be limited in the short to medium term by the high costs of capital and the lack of an enabling framework for such investments.

The paper consists of introduction and three sections. The first section provides an overview of the present state of financing water and sanitation infrastructure in the NIS. The second section presents a description and assessment of current planning and programming practices in the NIS water and environmental sectors. The role of financing strategies in the policy making is explained. The third section presents the methodological outline of financing strategies, summarises main findings and draws key conclusions and policy implications. At the end of this section, non-technical summaries of the three country studies are presented: Georgia, Moldova and the Novgorod Oblast of the Russian Federation.

2. Overview of financing water and sanitation infrastructure in the NIS

In the 1990s the financing of water and sanitation infrastructure in the NIS was characterised by the following features:

- Heavy reliance on shrinking public budgets rather than user charges for the cost of delivering services.

- Non-transparent and inefficient subsidies conveyed through low rates of household user charges, arrears, and non-cash forms of payments.

- Few disbursements of donor grants and IFI loans, notwithstanding 10 years of project identification and preparation.

- No commercial financing with the exception of (not always voluntary) arrangements involving the operation of municipal infrastructure by large enterprises.

Low funding of public water infrastructure can partly be attributed to economic recession and the overall resource constraints that this imposed on public budgets and investment. Economic recession has been an objective factor and a price in the transition to market-based economic systems. In 1998, the estimated aggregated real GDP in the NIS was only 53% of the pre-transition level of 1989. This can be compared with the recovery to 95% of pre-transition GDP in the countries of Central and Eastern Europe (Source: EBRD, Transition Report 1999). Some NIS, with per capita GDP below US$500 at PPP, belong to the poorest countries in the world.

2.1. *Financing by public budgets*

Little progress that has been made in most of NIS with tax reform, broadening the tax base or improving collection of government revenue (Himes, 1999). Budgets are small because governments do not collect enough tax revenue. Public sector revenue, as a percentage of the national income, remains very low by international standards, as shown in Table 1.

Table 1. **Shares of general government tax revenue in GDP in 1998**

Georgia	9.0%
Tajikistan	13.3%
Azerbaijan	15.3%
Kazakhstan	16.1%
Kyrgyzstan	17.6%
Armenia	17.9%
Ukraine	20.7%
Turkmenistan	26.3%
Belarus	27.2%
Uzbekistan	28.1%
Moldova	29.0%
Russia	31.7%
NIS average	**21.0%**
CEE average	**34.4%**
OECD average*	**36.6%**

* OECD data for 1998.
Source: EBRD 1999, OECD database.

The predictability of disposable public sector revenue is additionally distorted by the widespread practice in public budgets, budgetary organisations, funds and utilities of accepting non-cash forms of payments (barter, money surrogates). Central and local authorities and government-controlled entities further erode payment discipline through widespread use of tax and charge waivers, offsets and mutual settlements. The consequence of weak budget preparation was budget implementation flawed by ad hoc adjustments and non-transparent expenditure cuts undertaken during the course of a year. Reaction to emergencies and non transparent political bargaining are among the decisive factors driving the allocation of budget resources. Accumulating an unsustainable budget deficit was often a tempting way to avoid short-term problems (see Figure 1) and shift the burden to the future. In response to the consistently large public deficits the governments of most NIS have accumulated external and internal borrowing and payment arrears (to pensioners, public sector workers, budgetary entities, utilities). However, in the absence of economic growth, debt stocks have soared and are now substantial burdens on the economies; they have also considerably reduced creditworthiness of many regional and national governments in the NIS.

Control of public deficits involves increasing budgetary revenue, but also reductions in government spending. Water and sanitation infrastructure will most likely for several years suffer budgetary cuts like many other social services. However, as empirical studies on financing strategies have shown, there is room for rationalising budgetary expenditure cuts in the water sector by reducing operational subsidies and releasing resources for urgent capital investments. In different NIS, accumulated government obligations over more than a decade are likely to absorb almost all additional government revenue, even under optimistic assumptions about economic growth and future tax collection. These obligations include backlogs of spending on health service and education, payments of accumulated government internal obligations to pensioners

Figure 1. **Illustration of general government balances in CEE countries and the NIS**

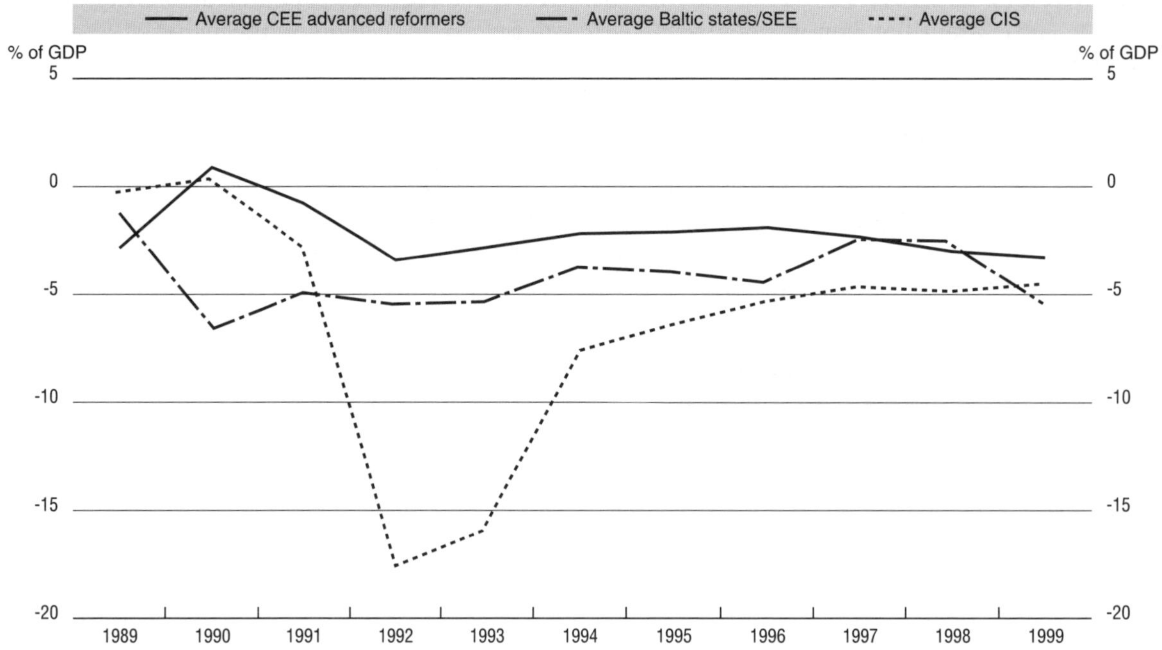

———— Average CEE advanced reformers   — - — Average Baltic states/SEE   ----- Average CIS

Source: EBRD Transition Report 1999. Data for 1999 are estimates.

and budgetary institutions, service of accumulated external debt, etc. Thus, a fundamental challenge to any effort to overcome the water sector crisis in the NIS is how to find realistic solutions under chronic conditions of scarce public funds and extremely limited capacity for additional borrowing.

Despite budget shortages, some funds were allocated annually to the authorities responsible for water and sanitation services. These funds however, have often been used in a manner that did not always maximise results in terms of the level and quality of services or environmental improvements.

- Funds available in the central budgets for capital, rehabilitation investments have often been thinly spread among too many beneficiaries and regions. This practice has not provided for the accumulation of a critical mass of financial resources necessary to prevent at least selected essential elements of water infrastructure from deterioration. Moreover, the main cities (in particular capitals) have attracted the bulk of central budget funds, leaving water infrastructure in the rest of the country in an increasingly critical condition.
- Budgetary funds have often de facto been used for operational subsidies.
- New capital investments have continued to be designed and implemented as turn-key projects by traditional design institutes, without adequate competition between consulting, design and contractor companies. Often this has led to infrastructure which is not cost-effective and which may not be appropriate in relation to the most critical service needs.
- Opportunities to leverage additional financing from public and private, domestic and foreign sources have been under-utilised. Lack of funding from domestic sources has not been compensated by an inflow of foreign funding. At the end of the first decade of transition, despite 10 years of identification and development of water and sanitation investments, only a few donor projects, financed by grants, and only a couple of projects financed by IFI loans have reached the investment phase.

2.2. Financing by retained earnings of water utilities

Central governments have been transferring responsibilities for the provision of water and sanitation services to the lower levels of governments, without the accompanying transfer of resources, or access to resources. Inappropriate tariff and subsidy policies, inappropriate definition of full cost in regulation which determines cost plus tariff calculations, inappropriate accounting practices, and inefficient billing and collection have all made it very difficult for water utilities to retain adequate earnings and become financially sustainable service providers. Water utilities sometimes do not have enough revenue to cover the operational costs of the system they operate and there is rarely enough to maintain assets adequately. This is partly due to the fact that the technologies they use are inefficient and excessively costly to maintain and operate. Emergency capital investments are provided from public budgets. In some regions, the misconception is widespread that water utilities – by virtue of being monopolies – continue to be rich organisations. This misconception effectively hinders policy improvements.

User charges for households in many countries are much below the levels that are considered affordable for low and medium income countries, using 4% of disposable income as a benchmark for what households can afford. Affordability is an important issue for the poorer segments of the population, but current subsidy schemes often support categories of people (*e.g.* war veterans) rather than households, irrespective of income.

2.3. Financing by international financing institutions (IFIs) and bilateral donors

IFI and donors provide financing for (non-investment) technical assistance and capital investments. The main value of the foreign public finance is its catalytic and demonstration effect. Commitments to capital investments usually encourage local governments and utilities to improve asset and finance management and to implement institutional reform programs. IFI loans are the only available source of long-term (15-20 years) debt. The cost of this debt is quite low for the lowest income countries, which are eligible for terms offered by IDA (International Development Agency) of the World Bank Group. IDA loans typically involve 10 years grace period and 40 years maturity. However, in volume terms the inflow of foreign public funds to water infrastructure in NIS has been modest reflecting serious institutional obstacles to effective demand for investments in the NIS water sector. IFIs also remain concerned about the short-term viability of water utilities under the present legislative and institutional set up.

2.4. Financing by the private sector

The banking sector in NIS is still undergoing transformation and consolidation. Lending activities of the banks mostly involve providing working capital, often for public budgets or state owned enterprises. Most bank loans are for less than one year. Longer investment credits come mainly from on-lending of IFI loans. The interest rates and spreads are still high and volatile reflecting the fragility of private financial markets. Only a few private strategic investors have shown any interest in NIS water utilities.

3. Strategic planning and programming in the water and sanitation sector

3.1. Strategic programming by the NIS governments

The NIS governments have made efforts to introduce a multi-year perspective and more strategic budgeting into public finance management. In most countries, this involved developing target-specific government programs. Target-specific government programs have particularly proliferated in the Russian Federation, Belarus and Ukraine. Almost every government agency has prepared several programs targeted at different problems in their area of responsibility, with lists of specific measures and investments to be undertaken over several years. The rough estimates of initial expenditure needs are attached to each item on a list, and desired shares to be covered by the different sources of financing (usually broken into central budget, local budgets, extra-budgetary funds, enterprise earnings and foreign sources) were identified. Line Ministries lobby for such programs to be officially approved at the government level. However, government approval does not translate into inclusion of related commitments in annual budgets. Most government commitments remain chronically under-funded. Available data from Ukraine reveal similar patterns.

Public investment programs (PIPs) are another instruments of multiyear investment planning. They were successfully used in the small Baltic countries, and can be effective mechanisms through which funding from the State (Republican) budget is allocated to public sector *investment* projects. They are usually managed by the Ministry of Economy and have the potential to introduce a rational, multi-year perspective into the allocation of scarce budgetary funds for long-term investments. PIPs are intended to streamline limited domestic and international resources and allocate them as effectively as possible to achieve maximum economic and social benefits, and to support the Government's long-term development strategies. Ideally, the planned government commitments under PIPs are effectively incorporated into the annual budgets. In the NIS, so far Kazakhstan has introduced a system for the development and management of a PIP.

3.2. *Environmental programs*

Sectoral planning tools, such as National Environmental Action Programmes (NEAPs) have been developed in most NIS. These programs were often valuable in identifying priority environmental problems. NEAPs often contain specific lists of the most urgent measures required, including investments to address emergency environmental issues in the near term. Some NEAPs (*e.g.* in Moldova) focus on plans to strengthen environmental monitoring, the regulatory framework and policy institutions.

NEAPs are not, and were usually not intended to be, implementation strategies to solve priority problems. In addition to identifying problems and list of urgent measures, NEAPs often include broad statements of environmental objectives for the country. What NEAPs do not include, however, is a set of SMART (Specific, Measurable, Agreed, Realistic and Time-bound) targets. Estimates of costs and expenditure requirements are not made (with the exception of occasional rough estimates of investment expenditures), plans are not realistically linked to finance, and affordability issues are not addressed. NEAPs were often too ambitious; they were generally based on the assumption that finance would be available for all capital investments needed to meet the targets, even if the cost of the action plan was unaffordable for the economy. This would be the case if the country (region) could not finance – out of current national (regional) income – the operating and maintenance costs of new (higher) levels of fixed assets plus the operating costs of the new institutions that needed to be put in place.

3.3. *Implementation and financing strategies*

Programs like NEAPs can be considered as important first steps towards sound strategic planning; in particular, they have often established a firm basis for achieving desired environmental objectives. They need to be complemented by realistic implementation strategies, including viable, long term, investment and financing strategies. Financing strategies can provide essential quantitative support and a reality check for the strategic implementation programs. To play their role effectively, financing strategies for the urban water sector should:

- Use inputs in the form of specific, measurable and time-bound targets, detailed quantitative data, and transparent assumptions.
- Contain robust estimates of costs and expenditure requirements, maintenance expenditure and operational expenditure for the system that is currently working, as well as investment and O&M expenditure for system rehabilitation and extension to meet more ambitious targets.
- Contain realistic, quantitative forecasts of finance available from different institutional sources and specify the form in which this finance can be provided (*e.g.* grant, debt or equity).
- Present cash flow forecasts for the programs and identify the financing gaps during the whole lifetime of a program.
- Identify policy and institutional options for closing the financing gaps by increasing the supply of finance, reducing costs or adjusting targets, taking into consideration what households, public budgets and the national economy can afford.
- Be supported by a computer model allowing easy, iterative simulations and scenario analyses.

Drawing *inter alia* from lessons learned from two previous environmental financing strategies for Lithuania and Armenia, a new methodology for financing strategies (FS) was developed by the consultants COWI, within the framework of the EAP Task Force, with the support of the Danish government. The methodology was applied and preliminary results were obtained for Georgia, Moldova and the Novgorod Oblast of the Russian Federation. Applications in Kazakhstan, Ukraine and Pskov Oblast in Russia are under way. At this stage the focus is on the municipal water supply, but it is envisaged that the methodology further elaborated to include other areas of environmental policy.

To support development of FS, an interactive spreadsheet model was prepared. The model requires input in the form of specific, quantitative policy targets and assumptions and forecasts about future changes of several economic and social variables. The model consists of three main modules:

1. the costing module, which calculates the demand for financing in terms of costs of meeting specific environmental targets.

2. the financing module, which quantifies the finance available from various public and private institutions in different forms (financial products).

3. the gaps calculation module.

Different levels of user charges can be assumed with due consideration of what is, and what will be, affordable to households. The financing module also includes rules that match specific sources and instruments of financing with specific project owners and project types. The model uses this information to calculate two interrelated gaps:

1. financing (or cash-flow) gap, which is the difference between needed expenditure and available financing, and

2. affordability gap, which is the difference between required annual cost and the realistic share of national income and public sector revenue that can be allocated to operation, maintenance and re-creation of needed assets.

The environmental finance strategy, *sensu stricto*, is elaborated by running the model several times, with different targets and different parameters representing various available packages of policies affecting demand and supply of financing. This interactive process of multiple model runs and policy assessment is needed to define realistic strategies. Realistic strategies are those for which all three gaps are closed with feasible policy packages. The financing gap can be bridged by a combination of increasing the supply of finance and decreasing demand for financing (*e.g.* through saving costs or revising targets). The affordability gap can be bridged only by a combination of the development of the economy and decreasing the demand for financing.

The niche for financing strategies in the sectoral policy cycle is illustrated in Figure 3.

Financing strategies can help to link feasibility studies at the project level with macroeconomic and budget planning, a linkage which is often not examined. They should not, however, be perceived as a substitute for one or the other. Although both municipalities and IFIs analyse affordability and the liquidity related to individual investment projects, financing strategies provide a framework for systematic aggregation of these and other projects on a regional and national level in order to assess their joint implications for domestic policies and budgets. This value added was clearly demonstrated in Georgia, where the World Bank is developing a project for rehabilitation of water and sanitation system in Tbilisi, while the European Commission is encouraging rehabilitation of the wastewater treatment plants along the Black Sea coast. Each party was making independent assumptions about the availability of co-financing from the central budget of Georgia, without full information of the aggregated claims on the consolidated budget. Merging these two ambitious investment programs, as well as other programs related to water services in other parts of Georgia, into the framework of a financing strategy helped identify, in quantitative terms, the difficult trade-offs that the Georgian budget planners would face if they wanted to fulfil all these commitments.

Figure 2. **Simplified structure of the model used to development of financing strategies**

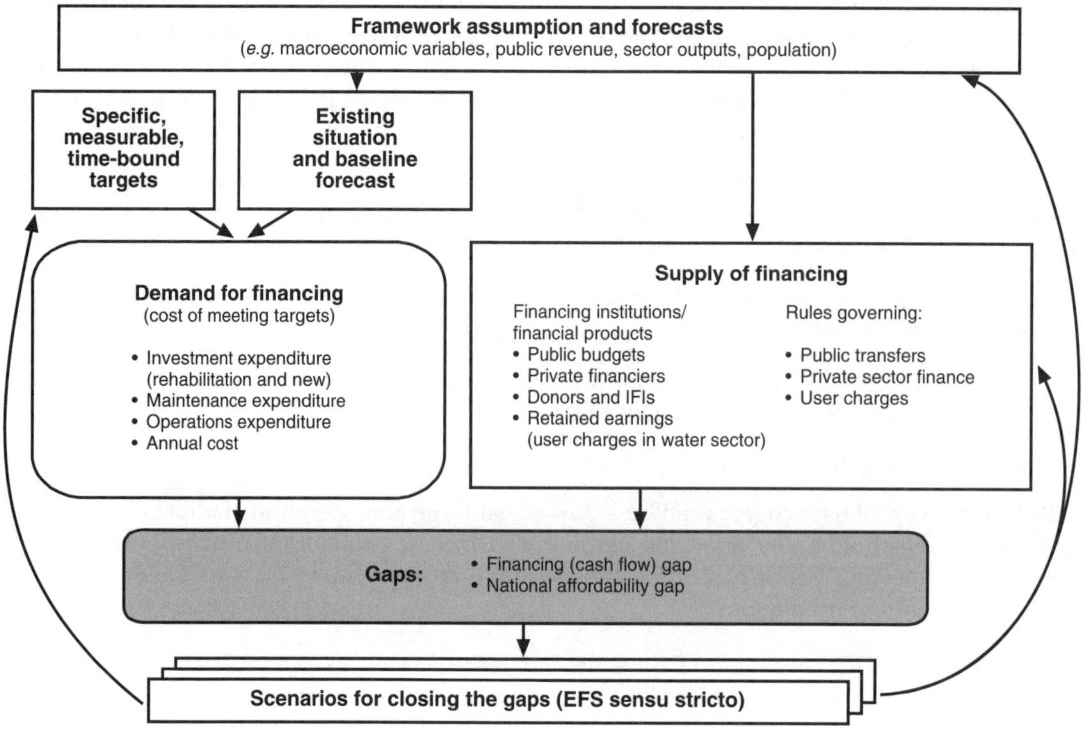

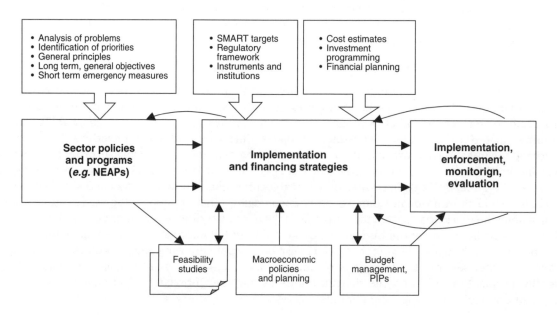

Figure 3. **Sectoral policy cycle and the role for financing strategies**

4. Results of the first financing strategies in the NIS

Development of the financing strategy begins from calculating the expenditure needed to move from the existing situation to a desired situation with respect to the levels and quality of water and wastewater services. However, the definition of the existing situation in the NIS water sector is not obvious. Built and installed infrastructure is not always actually used. Typically only a fraction of installed capacity (especially in wastewater treatment) is presently in operation. The remainder rests idle, gradually deteriorating. In the analysis for the finance strategies the term *existing situation* denotes the current level and quality of actually delivered water and sanitation services, and not their design performance.

The service level provided by a water and wastewater system is determined by its design, the value of assets and how it is operated. As systems grow older, repairs and replacement must take place in order to maintain designed service levels. As an international rule of thumb it is estimated that an equivalent of about 3% of the value of assets should be spent annually for maintenance (repairs, replacement of worn-out parts) in order to keep the service level unaffected. If repairs and replacement do not take place, service level will decline until one day the system would no longer be operational. The change in the assets' value of the system is used as an indication of whether the system is being "sufficiently" repaired/maintained. The maintenance backlog expresses the cumulated decline in assets' value. Although the relationship between assets' values and service level is indirect, the maintenance backlog is still a good indication of the development in service level. The larger the backlog, the larger the likely fall in service. It also indicates the amount of payments that are deferred, but that will have to be made in the future in order to restore the system to its present (not design) level of performance. The relationship between operational, maintenance, rehabilitation and extension expenditure is illustrated on figure 4.

Figure 4a. **Operations, maintenance, rehabilitation and extension of infrastructure**

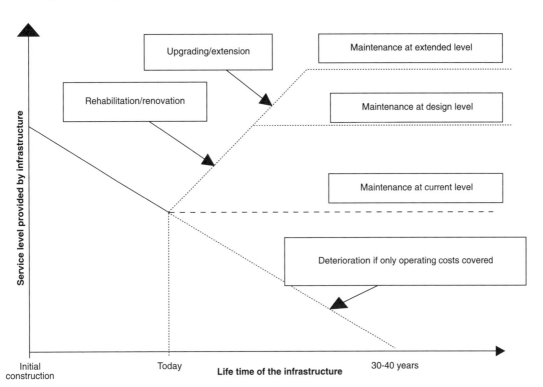

Each financing strategy begins by calculating the costs of preventing water and sanitation infrastructure from further deterioration. In other words, the baseline is to keep existing situation unchanged. The baseline includes the expenditure for operation and sustainable maintenance of the infrastructure, which is presently *in use*.

4.1. Key conclusions and policy implications from the empirical country studies

Preliminary results have confirmed, in quantitative terms, a serious crisis situation in the water and sanitation infrastructure in the NIS. The studies identified feasible, short-to-medium term actions to achieve realistic and affordable targets. The findings so far can be summarised as follows:

- Without increasing the supply of financing, NIS citizens should expect further deterioration of infrastructure and a decrease in the level and quality of services. This will entail *e.g.* discharge of untreated wastewater even from the largest cities, low pressures and interrupted supply of drinking water, and recurrent incidences of contamination of drinking water.

- Even to operate and maintain what still works now will require all three countries to implement demanding packages of policy and institutional reforms.

- While establishment of more ambitious targets of partial rehabilitation of infrastructure is possible (in addition to just operation and maintenance of what works now), the realistic achievement of these targets would require the countries to implement even more demanding packages of policy and institutional reforms to ensure additional finance. However, opportunities for additional finance will remain very limited for a long time.

- The necessary prerequisite for any realistic reform is to raise the water user charges for households. At present they are often established below the level of affordability (4% of household income). While charges for budget organisations and industries in most NIS are much higher, the collection rate typically is low. All other sources of finance taken together can not generate anywhere near as much resources for operations and maintenance. User charges will also be essential for attracting debt financing for rehabilitation investment.

- National and local budgets will have an essential role in the short and medium term in financing capital investments and facilitating access to credit. At present public funds are thinly distributed across the whole system. To achieve any improvement, however, the scarce funds of central budgets and donor grants will have to be narrowly concentrated (targeted) on rehabilitation of selected parts of infrastructure; some of the least important parts of the network may have to be left to deteriorate.

- IFI financing will play an important, though quantitatively limited role in the short term due to the low willingness to borrow for water infrastructure investments. Those governments that would be willing to borrow, however, may have their borrowing capacity constrained by high accumulated debt.

- Private financing can help bridge the financing gap for the large cities, but it carries a high price tag as well. Paying this high price is realistic only if increased costs are more than offset by efficiency gains and reductions in operational costs.

- The scope of projects under development by donors and IFIs needs to be carefully revised taking into consideration affordability of subsequent maintenance and operations. Care needs to be taken to ensure that financing commitments do not impose unrealistic claims on limited public and household budgets or compete with other priority environmental and social goals. Without realistic financing strategies, scarce public funds can be drained from other essential infrastructure in the country, accelerating the deterioration of social services elsewhere.

Experience to date has demonstrated that the financing strategy (FS) methodology can be a practical and powerful tool for policy planning and financial management. By explicitly addressing affordability constraints, FS forces unavoidable, though difficult discussions about priorities, trade-offs and cost-effective use of financial resources. FS can inform decisions about how scarce public funds can be used to mobilise financing from private and foreign sources. The development of effective FS requires policy makers and experts from the NIS to take "ownership" of the strategies and this, in turn facilitates the shift from a "needs mentality" to an "affordability mentality".

4.2. Case study: Georgia

4.2.1. Existing situation in water and sanitation services

About 95% of urban, and 35% of the rural population in Georgia is supplied by centralised water service. This indicates high network coverage by international standards. The actual performance of this system, however, is less impressive. Poor quality of the distribution network results in a rate of water loss between 10%-51% (40% in Tbilisi). All urban households suffer interrupted supply, receiving water much less than 24 hours a day, in some cities as little as 8-10 hours a day. This affects mainly people occupying higher floors of buildings, because of low pressure in the system. In rural areas the supply system often does not function at all. The major reason for this is the shortage of electricity supply. The majority of the connected urban households can have potentially good water quality as the main source is groundwater. Groundwater sources provide about 90% of the water supply apart from Tbilisi. (In Tbilisi 44% is from surface water). Drinking water quality problems are related to leaking pipes and cross contamination from the sewage system.

The centralised sewage systems exist in 37 towns in Georgia. 78% of population is connected to sewerage indicating high network penetration by international standards. The systems are, however, in poor condition. Wastewater treatment facilities were built for 33 towns, with the total daily, design capacity of 1.42 million m^3. There are 19 traditional mechanical/biological treatment plants, with a total design capacity of 1.39 million m^3/day and four purely mechanical treatment plants with a design capacity of 0.03 million m^3/day. However, the plants are typically 10-25 years old; some are as yet unfinished, and most are not maintained. None of existing plants is actually providing biological treatment since the technical facilities are out of order. Power and other resources are also needed. They are not delivered, because they are not paid for. Mechanical treatment is effective to a certain degree only in Tbilisi, Rustavi, Kutaisi, Tkibuli, Gori and Batumi with total estimated, daily capacity of 0.7 million m^3.

Preventing further deterioration of water supply and wastewater collection and treatment infrastructure means replacement of all assets actually used in 2000 according to the linear depreciation (deterioration) rate and provision of mechanical wastewater treatment in the five treatment plants that are in operation in 2000. Presently maintenance is insufficient, therefore both water supply and the wastewater systems deteriorate. Adequate maintenance would stop deterioration, but would not improve the system to its original, design characteristics. For the five plants that are designed for mechanical and biological treatment, but currently provide mechanical treatment only, maintenance means continuation of mechanical treatment only.

4.2.2. Available finance

The fiscal position of Georgian government is extremely weak. The share of government revenue in GDP is 9% – the lowest in the NIS. Ministry of Environment experts estimated that in 1999 about 1.2% of the government budget was spent on environment and water supply. Environmental expenditure survey indicates that about 80% of the total environmental spending is related to water and wastewater. The public budget contribution was calculated at the level of somewhat more than GEL 12 million in 2000 (US$6 million). These public subsides are meant primarily to compensate water utilities for the revenue foregone if the government exempts some users (*e.g.* pensioners) from the water charges and to cover a portion of operations costs if the government does not establish tariffs at the level that covers operational costs. A small share of government subsidies is available for maintenance work and rehabilitation investments, mobilised mainly in crisis situations. In future, budget subsidies are assumed to grow in line with national income (GDP).

Today, the average monthly household water bill (covering both drinking water and wastewater) constitutes only 0.8% of average household income and provides an estimated revenue of 17 million GEL (US$8.7 million) per year. The current collection rate is estimated at 40-60 % in recent years. However, a 70% collection rate was assumed for 2000, since local experts indicated that serious measures were planned. The revenue from industrial consumers and from budget organisations are estimated at 15 million GEL (US$7.7 million) per year and it is assumed that this source remain constant. A collection rate of 100% was reported by utility representatives. Payments in kind and through offsets by budgetary institutions were considered to be a problem until 1999, before a presidential decree prohibited them.

Expenditures needed to properly operate and maintain existing, low levels of service compared with the levels of funding currently available for water infrastructure are shown in Table 2.

According to the calculations, just trying to maintain the present infrastructure generates a financing gap through the next two decades if the major characteristics of financing do not change. If the present trends in the supply of finance continue there will not be enough money to provide even the present, low level of water and sanitation services. Even worse, under this "business as usual" scenario, it will be difficult to cover the costs of proper operation of existing systems. This conclusion is unaffected whether we assume lower growth or a more optimistic scenario of economic development. Moreover, the lack of annual maintenance accumulates resulting in accelerated deterioration of the infrastructure. Until revenues can be increased at least 3 times from their current 1999 levels, Georgian citizens should be prepared for continuing, further deterioration of infrastructure and decreasing level and quality of services. This will entail *e.g.* discharging untreated wastewater even from the largest cities, low pressure and interrupted supply of drinking water, lack of proper chlorination and recurrent incidences of contamination of drinking water.

The present level of available finance from all sources is enough to cover only about 27% of the total current costs of operations and maintenance of presently functioning water and sanitation infrastructure. User charges cover slightly more than a half of the costs of proper operations and only one fifth of the operations and sustainable maintenance.

Table 2. **Annual expenditure needs to properly operate and maintain existing water and sanitation infrastructure at 1999 levels of service and available finance under business as usual scenario in Georgia**

In million US$*

Georgia million 1999 US$*	2000	2005	2010	2015	2020
Baseline expenditure requirements					
Operational	28.9	28.9	28.9	28.9	28.9
Sustainable maintenance	52.7	52.7	52.7	52.7	52.7
Rehabilitation investment	0.0	0.0	0.0	0.0	0.0
New (extension) investment	0.0	0.0	0.0	0.0	0.0
Others (*e.g.* loan service)	0.0	0.0	0.0	0.0	0.0
Total baseline expenditure requirements in the water sector	**81.6**	**81.6**	**81.6**	**81.6**	**81.6**
Including water supply	53.6	53.6	53.6	53.6	53.6
Including wastewater	28.0	28.0	28.0	28.0	28.0
Baseline supply of finance					
State budget	3.2	4.6	6.6	9.7	13.8
Local budgets	2.7	3.6	5.6	8.2	11.7
User charges collected from households	8.6	8.6	8.6	8.6	8.6
User charges collected from enterprises and institutions	7.7	7.7	7.7	7.7	7.7
User charges total at current rates	16.2	16.2	16.2	16.2	16.2
International sources	0.0	0.0	0.0	0.0	0.0
Total financing available for the water sector	**22.0**	**24.5**	**28.1**	**33.7**	**41.8**
Baseline financing gap	59.5	57.0	53.4	47.8	39.6
Financing available as % of expenditure requirement	27%	30%	34%	41%	51%
User charges as % of operational expenditure	56%	56%	56%	56%	56%
User charges as % of O&M expenditure requirement	20%	20%	20%	20%	20%
Accumulated maintenance gap (backlog)	60	351	627	878	1 094

* 1999 constant prices and exchange rates.
Source: OECD.

4.2.3. Scenario – full maintenance of currently operated infrastructure

In this scenario the baseline target of merely maintaining the present service level is retained, but a package of measures are assumed to reduce and/or close the financing gap as soon as realistically possible, through increasing the supply of available finances. Five options are simulated.

- Increasing share of government budget in GDP from current 9% to 25% over 5 years, with spending structure remaining unchanged.
- Earmarking environmental taxes for water and wastewater spending (additional US$4.5 million p/a).
- Obtaining foreign grants in the amount of approximately US$1 million p/a until 2010 year.
- Utilising IDA loans (US$102 million, disbursed over 9 years, 35 years payback, annual interest rate 0.75%, grace period 10 years, 40% local co-financing required).
- Increasing households' user charge payments to 2% of average household income (increasing them to affordability ceiling of 4% would be difficult given the fact that the target does not provide for a significant improvement of the service level).

The results of simulations show that the single most important source of revenue comes from user charges. Even an increase to 2% of average households income (half of the internationally acknowledged affordability level) generates more revenue than all other sources taken together. Another important factor is the assumption concerning GDP growth. If these measures are implemented according to the assumed schedule, and the annual average growth rate in Georgia is 8%, it will be possible to restore the present level and quality of water and sanitation services in 20 years, although most parts of the system will continue to deteriorate in the short to medium term (up to 12 years). If growth is slower, the current financial deficit of the system will be persistent and the chances of restoring the present service level of the entire system by 2020 would be very small.

4.2.4. Scenario – rehabilitation of Tbilisi water supply

In this scenario, in addition to the maintenance targets, rehabilitation of pipes and pumps in Tbilisi water supply system was included. No rehabilitation of wastewater treatment system was envisaged

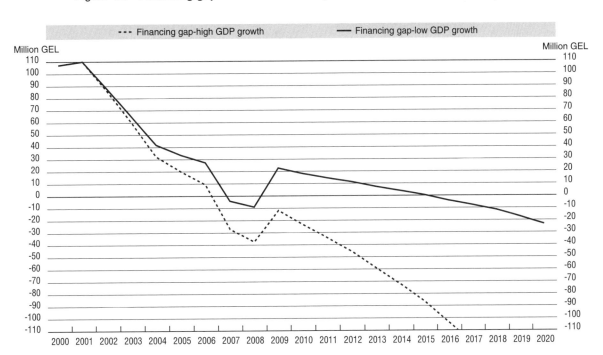

Figure 4b. **Financing gap for the scenario of partial rehabilitation of Tbilisi system**

however. With respect to the supply of finance, all assumptions are the same as in the maintenance scenario except for user charges. The rehabilitation results in a higher level of service, therefore it would be more feasible to require higher payments from households – 3% of average household income in Tbilisi and to 2% elsewhere. Metering of Tbilisi customers was also included. This would further reduce consumption and thus operational costs.

The results of simulation show that the increase of costs due to more ambitious targets in Tbilisi can be more than offset by increased revenue from higher user charges. Therefore the financing gap can be closed earlier than in the "Maintenance" scenario (by 2006 with high GDP growth). The financing gap would first decrease, but after 2008 would increases again due to the loan repayment. The date for closing the gap is very sensitive to the assumption about GDP growth. In case of low growth the financing gap could be closed only after 2014 and the system on average would not be restored to its present service level for the next two decades. This means that until 2014 investing in Tbilisi would tie up almost all available national finance and during that period infrastructure in other areas of the country would likely deteriorate further.

4.2.5. *Scenario - Rehabilitation of Tbilisi water supply and Black Sea Waste Water Treatment Plants* (WWTPs)

International obligations might put pressure on Georgia to reduce Black Sea pollution from municipal wastewater. In line with the actions planned in the NEAP, a scenario is analysed in which WWTPs in Black Sea costal towns are rehabilitated by 2020. All other assumptions on the cost and financing side are held the same as in the "Tbilisi rehabilitation" package.

Unfortunately, rehabilitation of infrastructure in several cities at the same time is going to be very difficult unless GDP follows the highest growth path. While local budgets will continue to spend their funds locally in a dispersed manner, difficult trade-offs will constrain the allocation of the central budget

Figure 5. **Financing gap for the scenario of rehabilitation of Tbilisi and the Black Sea coast WWTPs by 2020**

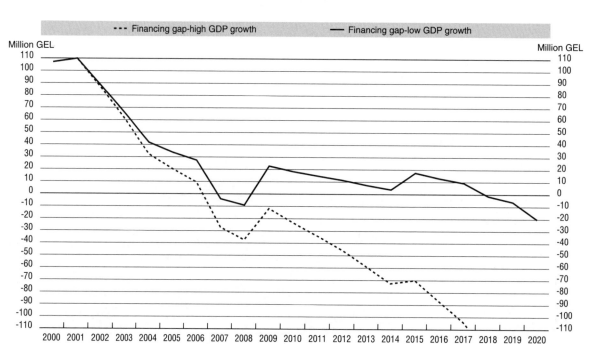

Figure 6. **Maintenance backlog for the scenario of rehabilitation of Tbilisi systems and the Black Sea coast WWTPs**

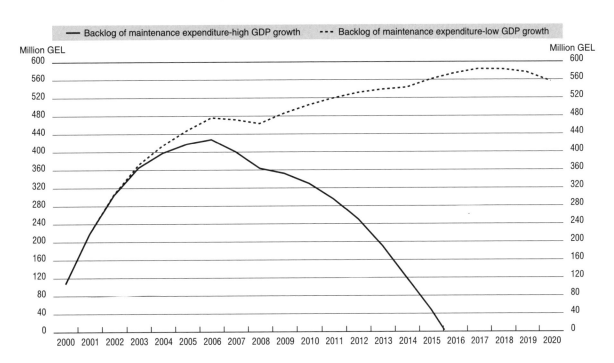

expenditure, foreign grants and loans. In the next decade, Georgia can afford only narrowly targeted priority rehabilitation investments. Given the very small size of the central budget for many years to come and limited borrowing capacity of the country, the funds controlled at the national level can be used either to effectively rehabilitate Tbilisi system or elsewhere (*e.g.* Kutaisi and the Black Sea cities) but not in all these regions at the same time. The crucial decision variable is the scale of rehabilitation of a Tbilisi system where maintenance currently accounts for about half of national costs. The water systems in the whole country can be brought back to their present performance within more than 16 years only under very optimistic assumptions about economic growth, as illustrated on Figure 6. Some accelerated regional investments may occur, if for example, the richest regions (such as wine producing counties that generate a bulk of government revenue) provide significant local contributions and make effective claims on the central budget.

4.3. *Case study: Moldova*

4.3.1. *Existing situation in water and sanitation services*

In Chisinau, Balti and Ungheni water supply is regular and generally of good quality. Towns between 25 000 and 50 000 inhabitants are regularly disconnected for 4-8 hours partly due to energy supply problems and due to the frequent breakdowns of worn-out pipes and pumps. Disconnection and lack of pressure in the network adversely impact water quality. Many smaller towns are supplied from groundwater sources which do not meet the chemical hygienic standards (problems with iron, Fluor, nitrate, pesticides etc.). These towns typically experience daily power cuts with resulting loss of regularity and water supply quality.

For towns larger than 25 000 inhabitants almost all wastewater is collected. For smaller towns, wastewater is typically only collected from the core of the town. The wastewater collected is generally led to wastewater treatment plants designed for mechanical and biological treatment. However, as a result of a combination of inadequate maintenance, power cuts, strong reductions in water inflow to treatment plants and limited financial resources for operation, most wastewater treatment plants designed for mechanical and biological treatment operate de facto with mechanical treatment only, if at all. At present, very few towns below 10 000 population treat their wastewater in any way. In the two largest cities, Chisinau and Balti mechanical and biological treatment is provided, although below its design efficiency.

4.3.2. Available finance

According to official statistics, financing of environmental expenditures today constitute 0.8% of GDP, most of this is for wastewater treatment, which is quite typical for low income countries. In addition 2.0% of GDP is spent on water supply.

The annual allocation from the public budget in 1999 was approximately MDL 15 million (about US$1.4 million). In Chisinau a water and wastewater improvement project is to be financed by EBRD and the state has committed to provide a grant of about 50% of the total investment expenditure. Thus, for the years 2000 to 2002, the public budget contribution is expected to increase more than three times, up to MDL 50 million (about US$4.8 million) annually.

Table 3. **Annual expenditure needs to properly operate and maintain existing water and sanitation infrastructure at 1999 levels of service and available finance under business as usual scenario in Moldova**
In million US$*

Moldova million of 1999 US$*	2000	2001	2002	2003	2004	2005	2010	2015	2020
Baseline expenditure requirements									
Operational	25.7	23.9	21.6	19.9	19.9	19.9	19.9	19.9	19.9
Sustainable maintenance	30.5	30.1	29.6	29.2	29.2	29.2	29.2	29.2	29.2
Rehabilitation investment	7.6	26.5	7.6	0.0	0.0	0.0	0.0	0.0	0.0
New (extension) investment	0	0	0	0	0	0	0	0	0
Loan service	7.0	6.9	9.3	9.1	9.0	8.8	2.7	0.0	0.0
Total expenditure required in the water sector	**70.8**	**87.3**	**68.2**	**58.3**	**58.1**	**57.9**	**51.8**	**49.1**	**49.1**
Including water supply	35.6								
Including wastewater	17.7								
Baseline supply of finance									
Public budget	4.6	4.6	4.6	1.5	1.6	1.7	2.4	3.0	3.8
Environmental Funds	0.4	0.4	0.4	0.4	0.5	0.5	0.8	0.9	1.1
User charges	25.1	25.1	25.1	25.1	25.1	25.1	25.1	25.1	25.1
Including households	8.8	8.8	8.8	8.8	8.8	8.8	8.8	8.8	8.8
Including other users	16.3	16.3	16.3	16.3	16.3	16.3	16.3	16.3	16.3
International grants	0.0	0.0	0.0	0.0	0.0	0.0	0.0	0.0	0.0
International loans	8.2	12.8	2.1	0.0	0.0	0.0	0.0	0.0	0.0
Total financing available for the water sector	**38.3**	**42.9**	**32.2**	**27.1**	**27.2**	**27.3**	**28.3**	**29.1**	**30.1**
Baseline financing gap	32.4	44.4	36.0	31.2	30.9	30.6	23.5	20.1	19.1
Financing available as % of expenditure required	54%	49%	47%	46%	47%	47%	55%	59%	61%
User charges as % of operational expenditure	98%	105%	116%	126%	126%	126%	126%	126%	126%
User charges as % of O&M expenditure required	45%	47%	49%	51%	51%	51%	51%	51%	51%
Accumulated maintenance gap	32	57	94	125	156	187	324	428	527

* 1999 constant prices and exchange rates.
Source: OECD.

With respect to user charges, on average the water bill is equivalent to 3.5% of average household incomes, which is close to the 4% affordability ceiling. However, since the collection rate is 65%, the actual payment by households is on average equivalent to 2.3% of average household incomes. There is considerable room for increasing user payments through increased collection rates. The inhabitants of Chisinau can afford more than people in other parts of the country. Moreover, all water utilities have a tariff structure which implies heavy cross subsidisation of households by industry and budget organisations. For example, in Balti, the tariff for industry and budget organisations is ten times the household tariff. The heavy cross subsidisation gives rise to problems of collecting cash revenues from industries and budget organisations.

Two foreign loans have recently been channelled to the sector. These loans have increased the supply of finance with a total of approx. MDL 600 million (US$57.75 million) over a five-year period. An EBRD loan (US$22.75 million) has mainly been channelled to rehabilitation of water supply and wastewater in Chisinau. In addition there is a limited component of "green-field" investments in the wastewater treatment. A Turkish loan (US$35 million) is targeted at rehabilitation, and limited "green-field" investments in extension of water supply in small towns and villages in the Southern part of the country.

Comparing expenditure needed just to prevent infrastructure from further deterioration and the currently available funding reveals significant financing gaps in the years to come. If the present trends in the supply of finance continue there will be not enough money to uphold even the present, low level of water and sanitation services. However, user charges – on average – are "just" enough to cover operating costs. In Chisinau current revenues exceeded operating expenditures, generating some operational surplus that can be used for some maintenance. The low level of service there seems to be attributed mainly to the physical disrepair of the system (accumulated maintenance gap). Outside Chisinau revenue are typically insufficient even to cover operating costs of water utilities. Paying for proper operations of what is in use may be a problem in these regions. In sum: if no more finance is provided, the existing, already deteriorated infrastructure will decay further and its operations will continue to be irregular and of a very low quality.

The present level of available finance is enough to cover over a half of the total current costs of operations and maintenance of presently functioning water and sanitation infrastructure. Over time this cost coverage would first decrease (rehabilitation expenditure and debt service) and than increase to 61% in 20 years from now, due to lower requirements for revenue if cost-caving investments are implemented in Chisinau.

4.3.3. "Users pay" scenario

This scenario brings significant increases in revenues from consumers, combined with cost savings and more realistic standards for wastewater treatment. With respect to cost saving measures, a package of measure has been assumed, such as replacement of pipes and pumps, extended coverage with meters, and water savings campaigns. The long term overall economic growth is assumed to be 5% per annum is assumed in this and subsequent scenarios.

Waste water discharge standards are assumed to comply with EU requirements of 15 mg BOD/l rather than with stricter, current Moldovan standards (3.5 mg BOD/l). This measure would not result in expenditure savings, however, because Moldovan standards are not (and realistically can never be) observed.

For user revenues, the following measures have been assumed:

- Increase collection rates for both household and non-household consumers to 85%.
- Increase household tariffs from 3.5% of average household income to 3.8%.
- Phase out cross subsidisation over a 20 year period.
- Increase share of cash payment to water utilities to 100%.

Figure 7. **Partial impact of increasing user charges up to affordability level on the backlog of maintenance**[1]

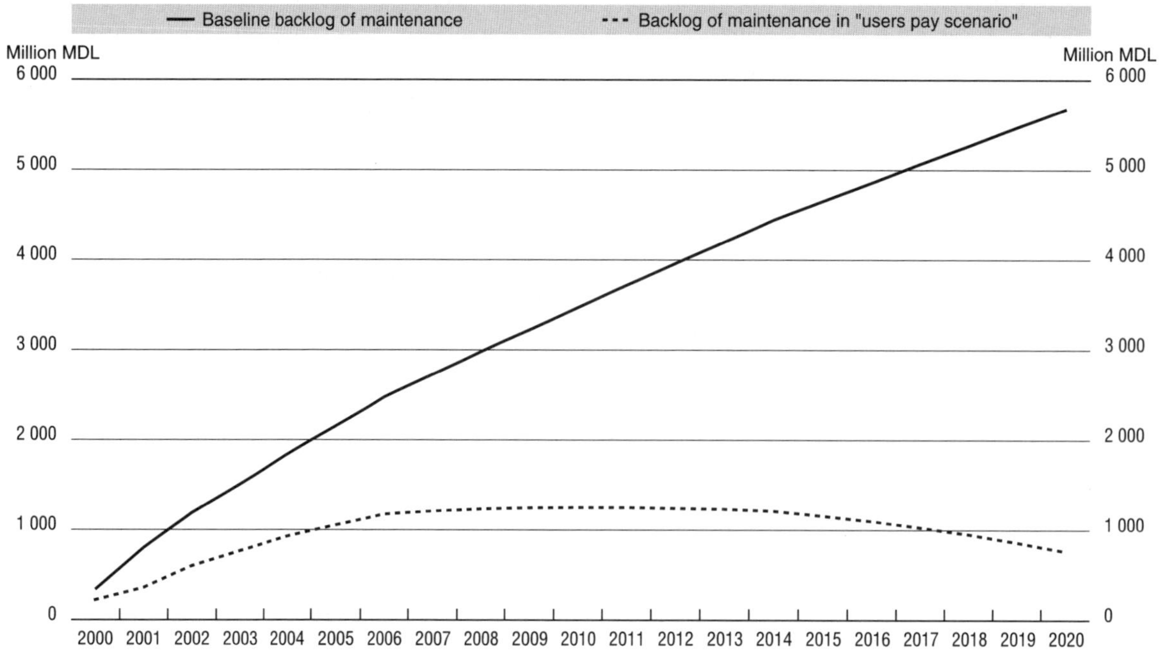

1. Upon the request of the Moldova authorities the reference scenario does include rehabilitation of some waste water treatment plants. Therefore it is not called "the baseline" on the graph.

Despite the challenging nature of these policy measures, they are not sufficient, by themselves, to achieve the target of maintaining the existing situation. Even with success of all the policies listed above water supply and wastewater infrastructure will continue to deteriorate for another 10-2 years. Only then will it be possible to renovate the system back to the current levels within another 15-20 years. To restore the present level of services more quickly, additional finance must be mobilised.

4.3.4. Additional finance scenario

This scenario combines the user pays scenario with additional public budget funding used to leverage concessional finance in the form of a US$62 million loan over five years.

Public expenditures for water sector could potentially increase to 2% of total public expenditures (net of debt service). Such a measure, although politically difficult to implement, in itself would not contribute significantly to reducing the financing gap as illustrated on the Figure 8, below.

However, these budgetary resources may be used to leverage additional concessional loan on IDA terms. Therefore, concessional loan of MLD 651 million (US$62 million), disbursed over the five year period 2003-2007 was simulated. The impact of a loan on restoring the present level of service is more significant; however, by itself, it can not close the financing gaps either.

Figure 10 below illustrates the combined impact of all measures listed above on the maintenance backlog. It shows that only a combination of all the actions assumed in the "additional financing scenario" is sufficient – and necessary – to close the gap. It would still take 5 years to close the current financing gap mainly due to the time it takes for the policy measures to have effect. The maintenance backlog can

Figure 8. **Partial impact of additional public finance on the backlog of maintenance**

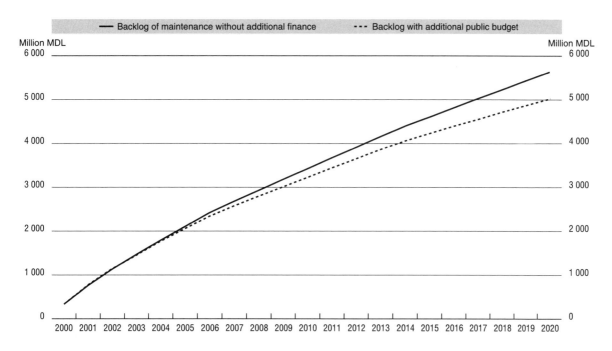

Figure 9. **Partial impact of US$62 million concessional loan on the backlog of maintenance**

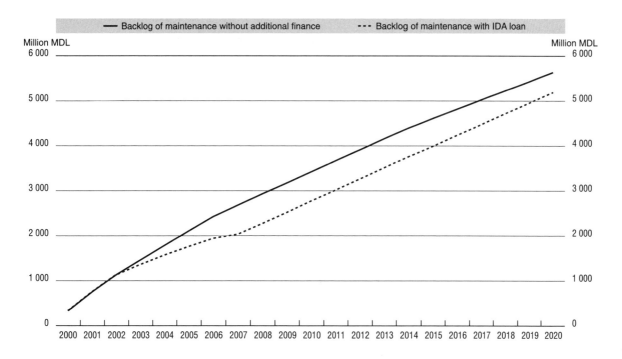

Figure 10. **Combined impact of all additional finance measures on the backlog of maintenance**

― Backlog of maintenance without additional finance   --- Backlog of maintenance in "additional finance scenario"

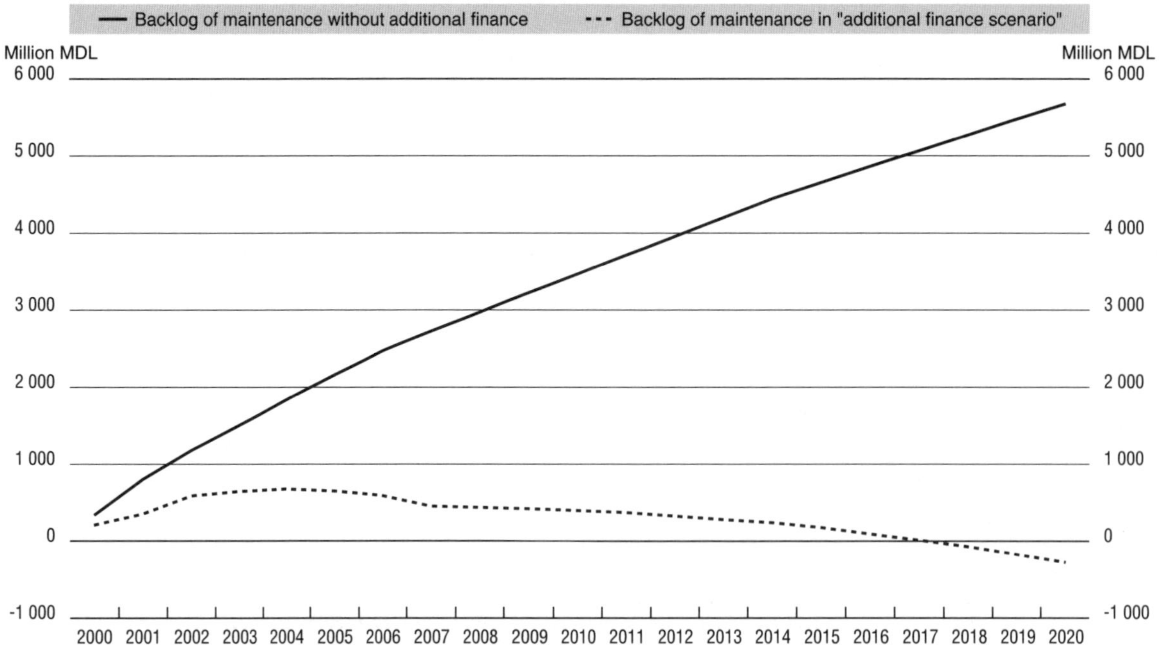

then be removed, and the present service level restored in approximately 13 years. Only after, would Moldova be able to afford additional rehabilitation investments improving the performance of water and sanitation systems to their design parameters. The single most important revenue source is user charges. Raising them only moderately within the affordability limit, and improving collection rates in cash, has the potential to generate more money than all other sources taken together.

Table 4. **Policies and measures sufficient to close the financing gap and restore the present level of service within less than 20 years**

	Users pay scenario				
In fixed 1999 US$	2000	2005	2010	2015	2020
Initial financing gap	30.6	28.7	21.6	18.2	17.1
Energy savings	0.0	1.4	4.3	4.3	4.3
Total additional user charges of which:	6.0	11.4	13.0	15.3	18.8
Additional households	3.0	6.0	10.0	15.2	21.9
Additional non-household	3.0	5.4	3.0	0.1	−3.1
Additional foreign grants	3.1	3.1	3.1	3.1	3.1
Resulting gap	21.4	12.7	1.1	−4.6	−9.0
	Additional finance scenario				
In fixed 1999 US$	2000	2005	2007	2015	2020
Additional public budget	0.0	2.0	2.6	3.8	4.9
IDA loan	0.0	12.5	15.5	0.0	0.0
Loan service	0.0	0.0	0.0	−1.9	−3.6
New financing gap	21.4	−1.8	−17.0	−6.5	−10.3
Backlog with additional financing	21	63	43	17	−26

Source: OECD.

4.4. Case study: Novgorod

4.4.1. Existing situation in water and sanitation services

The coverage of centralised water supply systems in the main cities and urban settlements of the Novgorod oblast, as measured by the share of the population actually served, ranges from 99% in Novgorod city to about 70% in small cities and in smaller town settlements. The coverage includes centralised water system as well as water stands; the latter may serve about 2-3% of the population in Novgorod and up to 25% of the population in smaller urban settlements.

The highest regularity of water supply, measured as the share of the connected population covered by centralised water supply enjoying 24 hours service, is recorded in Novgorod (98.5%), and exceeds 90% in all cities and towns. Irregularity is mainly due to technical problems in some high-rise blocks (4-5 storey buildings with insufficient water pressure from the pumping station).

Surface water sources account for more than 80% of raw water supply. Drinking water often does not comply with the national GOST standards in terms of colour, Fe, residual Al and chlorine concentrations at the tap, mainly due to the poor state of the distribution network. The degradation of the water supply network causes secondary pollution. Intensive treatment with chlorine also has an adverse impact on human health.

The municipal wastewater treatment facilities in the 10 major cities all have mechanical-biological treatment facilities installed. Out of the 11 small urban settlements for which information on municipal wastewater treatment facilities was obtained, 4 cities had mechanical-biological treatment facilities, 4 cities had mechanical treatment facilities, and 3 cities had no treatment facilities. Installed treatment capacities are in general considered sufficient. Still, the poor state of the facilities, owing to the lack of funds for rehabilitation and under-investment, has resulted in substantial discharges of polluted wastewater. An additional 30 000 m^3/day of storm water and wastewater is discharged without any treatment due to insufficient wastewater sewerage systems. The poor state of the sewerage networks, in combination with the lack of storm water networks, results in infiltration. This causes high pumping costs and dilution of waters into the wastewater treatment plants that reduces the efficiency of mechanical-biological treatment.

4.4.2. Available finance

Public revenues reached a low level of 11% of gross regional product in 1998 reflecting the public finance and tax collection crisis. About 5% or RUR 90 million (US$ 3.6 million) of the consolidated public expenditure budget in 1999 was used for capital investments including those related to water. Less than 20% of capital expenditures in 1999, corresponding to 1% of the public sector revenues, was destined for the waste and wastewater sector. The Oblast administration expects that this share will remain unchanged in the medium term. Moreover, it has been assumed that GRP (Gross Regional Product) growth remains stable at 3% p.a. over the entire forecasted period.

The responsibilities for financing new municipal infrastructure development is assigned to the public budgets, while the responsibility for asset renewal and maintenance rests with the operating utilities and financing is provided through tariffs. However, almost 22% or close to RUR 460 million (US$18.5 million) of the consolidated public budget is destined for operational subsidies to the housing maintenance and municipal service sector. Of this, about RUR 90 million (US$3.6 million) was extended to the water and wastewater sector as operating subsidy in 1999.

The environmental funds system in Novgorod is relatively small in terms of revenue generation and the priority expenditures are non-wage related operating subsidies to the environmental administration. Total expenditures in 1999 amounted to less than RUR 6 million (US$240 000), of which only about RUR 1 million (US$ 40 000) was destined for environmental protection projects. The rest was used for administration support and pollution charge offsets. All disbursements were in the form of grants. In the baseline projection it has been calculated that under optimistic assumptions, the annual disposable resources of the environmental funds may be RUR 8-13 million (US$320 000-520 000) in the long term (years 2005-2010). However, as a result of prevailing spending priorities and the pollution charge offset scheme, only RUR 2-3 million (US$80 000-120 000) annually in real terms could be available for co-financing investment projects.

The financial sector in Russia remains subdued after widespread banking failures in 1998. The ratio between domestic credit provided by local banks and GRP is less than 2%, very low compared with developed market economies and even in comparison with economies with similar income levels per capita. In addition, the asset structure reflects that the banking sector is oriented primarily towards financing running costs of the government. Interest rates and interest spreads appear prohibitively high. Long-term investment credit is very scarce. Following the crisis in 1998 several laws were enacted on the restructuring of credit institutions and on the insolvency of financial institutions. However, according to EBRD (1999), restructuring has been "slow, uncoordinated and inefficient". It is not considered realistic within the foreseeable future that the domestic banking system will be able or willing to contribute to the long-term financing of infrastructure.

A number of international donors have supported environmental projects in the Novgorod region. DANCEE provided support to three investment projects in 1995-1999. Two projects concerned the rehabilitation and expansion of the biological WWT facilities in Novgorod (total cost about US$2 million). Finland has provided technical assistance and a small investment support (US$0.5 million). Sweden and USAID provided only technical assistance. Local co-financing in the order of 30% of investment costs is typically a minimum requirement.

Table 5. **Annual expenditure needs to properly operate and maintain existing water and sanitation infrastructure at 1999 levels of service and available finance under business as usual scenario in Novgorod**

In million 1999 US$

Novgorod, million of 1999 US$	2000	2001	2002	2003	2004	2005	2010	2015	2020
Baseline expenditure requirements									
Operational	6.5	6.5	6.5	6.5	6.5	6.5	6.5	6.5	6.5
Sustainable maintenance	17.4	17.4	17.4	17.4	17.4	17.4	17.4	17.4	17.4
Rehabilitation investment	0.0	0.0	0.0	0.0	0.0	0.0	0	0	0
New (extension) investment	0.0	0.0	0.0	0.0	0.0	0.0	0	0	0
Loan service	0.0	0.0	0.0	0.0	0.0	0.0	0	0	0
Total expenditure required in the water sector	23.9	23.9	23.9	23.9	23.9	23.9	23.9	23.9	23.9
Baseline supply of finance									
Public budget	5.5	6.2	6.9	7.1	7.3	7.5	8.7	10.1	11.7
Environmental Funds	0.1	0.1	0.1	0.1	0.1	0.1	0.1	0.1	0.1
User charges	6.2	6.2	6.2	6.2	6.2	6.2	6.2	6.2	6.2
Including households	2.2	2.2	2.2	2.2	2.2	2.2	2.2	2.2	2.2
Including other users	4.0	4.0	4.0	4.0	4.0	4.0	4.0	4.0	4.0
International grants	0	0	0	0	0	0	0	0	0
International loans	0.0	0.0	0.0	0.0	0.0	0.0	0.0	0.0	0.0
Total financing available for the water sector	11.8	12.5	13.2	13.4	13.6	13.8	15.0	16.4	18.0
Baseline financing gap	12.1	11.4	10.7	10.5	10.3	10.1	8.9	7.5	5.9
Financing available as % of expenditure requirement	49%	52%	55%	56%	56%	57%	62%	68%	75%
User charges as % of operational expenditure	95%	95%	95%	95%	95%	95%	95%	95%	95%
User charges as % of O&M expenditure requirement	26%	26%	26%	26%	26%	26%	26%	26%	26%
Backlog	12.1	23.5	34.2	44.7	55.0	65.1	112.2	152.8	185.8

Source: OECD.

Payments for water and wastewater as a percentage of household income is very low compared with international levels. On average, each household spent 0.4% of disposable income on water and wastewater services in 1999, almost ten times lower than in Moldova, which also has lower per capita income. Tariffs vary among water utilities. Industrial tariffs are 3-6 times higher than domestic tariffs and the region in general still relies on cross-subsidisation in order to keep the households tariff low. The present collection rates are estimated to be above 80% and approaching 90% for residential customers.

As indicated in the table below, maintaining and preserving the present infrastructure implies that total operations and maintenance expenditure would have to increase more than two-fold.

4.3.3. *Scenario of maintaining existing infrastructure*

Because the baseline gap is generated mainly by O&M expenditure requirements, it can be closed primarily by increasing revenue flow from user charges. A major re-balancing of present tariff policies would clearly be difficult to implement in the short term. Hence, it has been assumed that both the domestic and non-domestic tariffs gradually increase towards the levels that recover O&M costs in 2005. The average operation and maintenance expenditure for water and wastewater operations per m^3 of water consumed would need to increase on average five times, from its present level of about RUR 2 to approximately RUR 10 (US$0.35). This should be compared to the present domestic tariffs of RUR 1-1.50 per m^3 and the tariffs for budget organisations and enterprises of RUR 4-7 per m^3.

For domestic customers a move to full cost recovery in the water sector, would imply that the real tariff would have to increase from 0.4% of average household income in 1999 to 3.7% of income in 2005. This is considered to be within affordable limits (4% of average household income). Still the envisaged tariff adjustment might require targeted transfers from the public budget to households with low income. Affordability of household water bills is sensitive to projected household income levels. If real GRP and real salary growth declines to 1% per year from the baseline level of 3-4% per year over the period 2000-2005, the average water and wastewater bill will amount to more than 4% of average household income in the year 2005.

With projected stable macroeconomic conditions, it would be feasible for the industrial real tariffs to follow the average real costs of providing the services over the medium term. Tariffs for industry and budget organisations will have to double over the period 2000-2005. However, the increased cost burden can be partly offset by reduced water demand.

4.4.4. *Scenario of improving quality and level of services*:

Despite the difficulties of maintaining present service levels, Novgorod Oblast authorities and experts have identified a number of ambitious policy objectives related to water service quality and environmental standards. With the assistance of consultants, these objectives were translated into specific targets in terms of water quality at the tap, regularity of supply, wastewater treatment standards and the coverage of water and sanitation services. Specific target values for each city and group of cities can be found in the background consultants report.

As a result of the substantial tariff increases over the period 2000-2005, it has been assumed that water production decreases to 305 l/c/d. Wastewater discharges from households and industries decrease proportion to water demand. For industry and budget organisations, it is assumed that consumption declines by 25% from its present levels as the tariff is adjusted upwards by a factor of 2 in real terms over the period 2000-2005.

The substantial investments in water treatment would involve mainly substitution of surface water by ground water and renovation of water treatment facilities. Network rehabilitation will be essential in addressing secondary pollution in highly deteriorated water distribution networks. The extent of the renovation corresponds to replacing about 35-40% of the water distribution network over a ten-year period. In combination with the expected reduction in water demand this would reduce water losses in the water distribution system by at least 30%.

The targeted wastewater treatment technology for all cities is mechanical-biological treatment (MB), except for cities with less than 10 000 inhabitants, where mechanical treatment is assumed to be

Figure 11. **Financing gap under the baseline and the scenario of achieving targets of improvement in quality and level of services (in US$ million)**

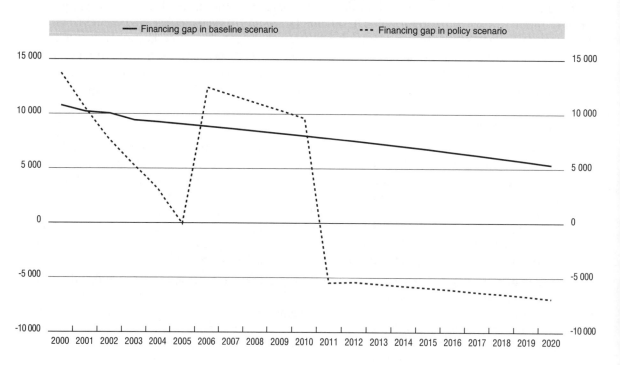

Figure 12. **Backlog of maintenance under the baseline and the scenario of achieving targets of improvement in quality and level of services (in US$ million)**

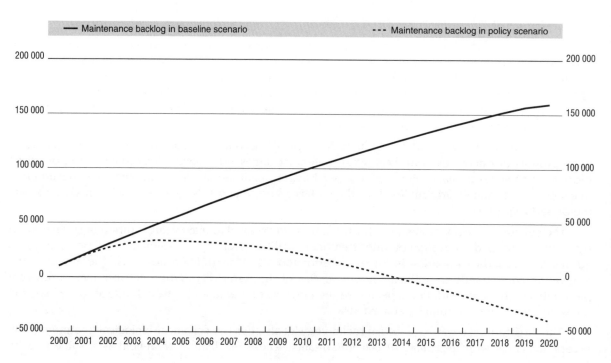

sufficient. The substantial investments in wastewater treatment would involve mainly rehabilitation of existing WWTP, so that they can provide efficient MB treatment. For all cities, the target renovation rate for the existing WWTP is 10%. For all the urban areas included in the strategy, 15-20% of the sewerage network is assumed to be renovated, while extension of the network will not exceed 2%.

Compared to the baseline scenario, achievement of these ambitious targets would require significant investment expenditure both in large-scale system renovation and even extension. New capital investments in urban water sector would begin in 2006, because this year the maintenance backlog is expected to decline to zero if measures envisaged in the maintenance scenario are implemented. Operations and maintenance expenditure would also increase due to the higher cost of maintaining renovated and upgraded assets. Later, however, they are expected to decrease due to a 25% operational cost savings. In order to achieve these targets additional financing would have to be mobilised. The package of policies to mobilise the additional supply of money for closing the financing gap is challenging.

1. Tariffs would need to increase gradually to full O&M cost recovery in 2005.
2. For non-household users, tariff collection rate as well as share of cash will need to gradually increase to 95%-100%.
3. Local funding will need to leverage foreign grants by a factor of 1, *i.e.* for each RUR in local financial contribution, the region can attract RUR 1 in co-financing. Foreign grants can decline to zero by 2005.
4. Loans will be required to implement the investment program. The loans will need to be disbursed in the period 2006-2010 in almost equal instalments of around US$ 10 million per year.
5. The consolidated public budget revenue will need to grow from present 12% to 15% of GRP in 2004, while public expenditure on the water sector will go down from 10% in 1999 to 5% of GRP in 2006 reflecting the fact that increased user charges will substitute operational subsidies from public budget.
6. Capital investments in water infrastructure will need to increase from 0.8-0.9% to 4% of the annual public budget expenditure. Even in that event, it would constitute less than 1% of GRP.

If any of these measures is not achievable, they would have to be compensated by a more aggressive application of other measures. The alternative would be to revise the level of targets and decide on a more modest rehabilitation program.

Chapter 4

BACKGROUND PAPER ON REFORM OF URBAN WATER SUPPLY AND SANITATION SERVICES

Introduction

This paper provides background on the problems facing the urban water supply and sanitation sector in the NIS region and the types of reforms that are required.[1] While some reform initiatives are underway, there are many signs of a crisis in the water systems of the NIS and urgent reform measures are needed to reverse the trend. Recent studies have documented critical problems, including poorly maintained infrastructure, frequent interruptions in service, high consumption relative to other countries, deteriorating water quality, and considerable wastage. If co-ordinated actions are not taken to modernise and reform the sector, it will continue to have adverse impacts on human health, productivity and important ecosystems, with negative impacts increasing over time.

The situation calls for short-term actions to address urgent needs, as well as a more strategic long-term vision of how the sector should operate in the coming decades. This paper highlights many of the issues that need to be addressed before the sector can be advanced, in terms of improved operating efficiency and investment financing. Obstacles vary from country to country, and the observations discussed do not apply uniformly to all countries in the region. The issues presented do, however, illustrate where targeted actions and new ways of working are warranted.

Although investments are required, the problems will not be solved by providing additional resources from the public budget: such an approach would not be sustainable and would reinforce the inefficiency of existing arrangements. In any case, the financial resources required to address the scale of the present problem simply do not exist. In some countries only 30-40% of the resources needed to operate and maintain the existing networks are available, and external financial resources could only help to address a small fraction of the total needs. While a number of international organisations and funding agencies are supporting reform projects in the NIS, their efforts are often impeded by existing policy, institutional and financial obstacles in project countries.

Experience gained since 1991 clearly indicates that a new framework is needed to guide the reform of the urban water supply and sanitation sector. Such a framework is essential to stop the continued deterioration and eventual collapse of water and sanitation services, with the serious consequences for the well-being of the population and their environments which this would entail. The objective of the reform should be to ensure that good quality water is delivered reliably and at least cost to the population. Some of the key reforms needed to achieve this objective include:

- Decentralising responsibility for water service provision from national authorities to the local level.
- Reforming water utilities (vodokanals) so that they have the autonomy, capacity and means to provide water and sanitation services efficiently and effectively, subject to strict supervision by public authorities.
- Engaging the public directly in the reform process.
- Establishing the sector on a financially sustainable basis so that funds are available to make necessary investments, while addressing the needs of poor and vulnerable households.

- Creating incentives to substantially increase efficiency in the use of water by consumers and in the operation of vodokanals.

Other specific reform recommendations are summarised in the Guiding Principles for Reform of the Urban Water Supply and Sanitation Sector in the NIS (CCNM/ENV/EAP/MIN(2000)6).

1. Signs of a Crisis

Despite the high service coverage of water supply systems in NIS urban areas, the poor state of facilities, lack of adequate maintenance, and insufficient resources available for operations, make the reliability and safety of the service a major concern. Water supply losses of 50% or more are common throughout the NIS region, which has been evidenced by a number of recent studies (see Box 1). There can be considerable variation in supply levels among different users; inadequate water supply within cities is often attributable to poor maintenance of pumps (or lack of funds to pay for the electricity to run them), particularly in high-rise buildings. In addition to lack of maintenance, poor technical design and use of poor quality materials in construction are frequently problems.

Box 1. Water supply losses in the NIS

- In Tbilisi, Georgia, it is estimated that as little as one quarter of supply actually reaches the consumer and up to 700 km of pipeline has seriously deteriorated.
- In Zaporizhzhia, Ukraine, 25% losses are recorded, but metering is not widespread and real losses are estimated to be closer to 50%.
- In Dushanbe, Tajikistan, losses are estimated to be as high as 70%.
- In the Aral-Sarybulak system in Kazakhstan (which serves 60 000 people), it is estimated that as much as 50% of the system needs to be urgently replaced due to excessive supply losses.

Source: *Obstacles and Opportunities to Commercialising Urban Water Services in the New Independent States* (NIS): *Final Report* by Environmental Resources Management (ERM) under contract to the UK Department for International Development (June 2000).

The region's wastewater treatment facilities are typified by a lack of treatment, or a limitation to primary treatment alone. Even where systems provide secondary treatment, poor maintenance, high electricity costs and limited financing often reduce system efficiency.

The deterioration of water supply and wastewater treatment facilities across the NIS has meant that some health problems are now more akin to those of developing countries, namely a prevalence of diarrheal diseases and gastrointestinal illnesses associated with unsafe water. For example, in recent years, epidemics of typhoid and dysentery have occurred in various parts of Russia, as well as cholera outbreaks in Moscow and other large cities. Infant mortality caused by intestinal infectious diseases is particularly high in Central Asia.

Water quality problems are also linked to colour, taste, odour, and chemical/bacteriological contamination. Quantifying the water pollution problem in the NIS is not easy – few countries have fully operational national water quality monitoring programmes, and so water quality and pollution data tend to be project-specific and anecdotal in nature. However, it is generally accepted that micro-biological contaminants are present in drinking water throughout the region – health statistics alone support this argument, in addition to the many studies which have been carried out as part of urban water services development projects. Water quality problems also extend to industrial and agricultural pollutants.

Aside from the considerable health impacts of water pollution, there is an impact on the region's economic productivity due to premature deaths and workdays loss to illness. See the background paper on "Valuing Environmental Benefits and Damages in the NIS: Opportunities for Integrating Environmental Concerns in Policy and Investment Decisions" (CCNM/ENV/EAP/MIN(2000)3) for more information.

The socio-economic impact of poor water supply in the NIS has been studied to some extent, particularly as an adjunct to the various water utility reform projects supported by international donors. Generally, the studies confirm that the groups most affected by poor water supplies tend to be women (who usually bear responsibility for securing alternative supplies); the elderly; those living on higher floors of apartment blocks; and the residents of peripheral communities. One study found that low-income households spend a significantly higher percentage of their household income on coping strategies than wealthier families, including a variety of water purification methods. Households with higher incomes and higher levels of education also more actively seek safe solutions to their water supply problems.

Under the Soviet system, water and sanitation services were provided to domestic customers at very low prices, and the government, which controlled a very large share of the national income, provided subsidies in various forms (either as direct transfers or through charging higher tariffs to industries and state budget organisations). The government also funded most capital improvements. With the transition to a market economy, the subsidy levels have in many cases decreased, but tariff levels have remained significantly below cost-recovery levels, causing an overall state of financial crisis in the sector. This effects both the ability to maintain existing service levels, and the ability to make necessary investments to rehabilitate, upgrade and expand the systems.

2. *Planning for Urban Water Sector Reform*

The objective of the reform should be to ensure that good quality water is delivered reliably and at least cost to the population. To achieve this goal, the level of co-operation among different institutions must be increased to promote discussions on long-term strategy and realistic policy targets. An objective-setting process should allow for the participation of all relevant stakeholders, in order to build a diverse support base for upcoming changes. Careful choices and priority setting will be essential to obtain the largest benefits from invested resources, and all targets should include an realistic appraisal of financing needs.

Figure 1 depicts the various governmental and non-governmental stakeholders currently involved in a typical arrangement for the delivery of urban water services in the NIS. The national authorities should facilitate a participatory planning process involving all these relevant stakeholders. Strategic goals for the water sector should be laid out in a sector strategy or policy paper, which should be complemented by an action and financing plan (see the background paper on "Financing Strategies for the Urban Water Sector in the NIS" (CCNM/ENV/EAP/MIN(2000)4). The process should identify performance and quality goals based on available resources, and targets should be set only after alternative scenarios are evaluated in terms of costs and benefits. Box 2 outlines a possible structure for a Water Sector Reform Group at the national level. Once a long-term strategy has been developed, short and medium-term actions can be elaborated for all stakeholders, and detailed implementation timetables and financing plans prepared.

Many of the tough reforms needed in the water supply and sanitation sector are not taken due to concern about impact on the poor. However, the poor are affected proportionally more than others by lack of services and end up paying higher prices for alternative sources of water. Affordability is mentioned as the main reason for maintaining domestic tariffs at a low level, yet low tariff revenue is one of the key factors in the system's deterioration which forces the poor to spend a significant portion of income coping with the deficient services. Currently, they do not have a voice in the planning of the sector and key decisions are taken without consultation. The reform process should take account of the needs of the poor in all planning stages.

Figure 1. **Typical NIS Arrangement for urban water services delivery**

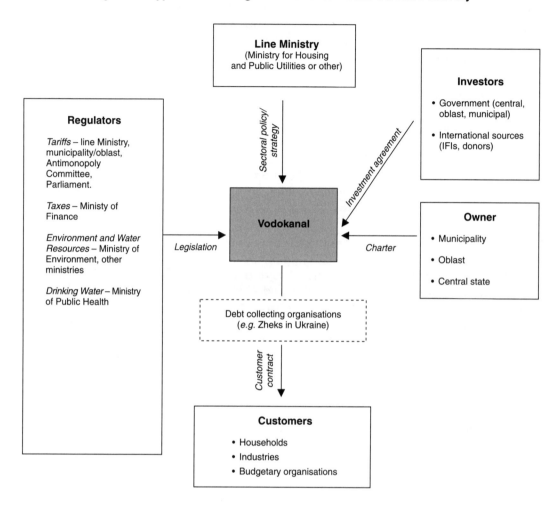

At the local level, service planning should be accompanied by socio-economic and willingness-to-pay studies, to reveal what level of service households are willing and able to support, as well as provide information on the needs of poor and vulnerable households.

3. Institutional Arrangements

There are typically many institutions involved in the water services sector of the NIS. Often, their exact roles and responsibilities, as well as their mandates and legal powers, are not entirely clear. The result is an ineffective institutional framework that complicates utility regulation, strategic planning and development of a common vision for the future. Day-to-day management of water utilities is complicated by the fact that many external parties play a role in deciding on tariffs, investments, and other matters, delaying the decision-making process for substantial lengths of time, and potentially subjecting it to political interference.

This can lead to "common agency" problems, which occurs when several agencies impact on the operating conditions of a firm, (in this case the water utility), have insufficient co-ordination, and this results in conflicting objectives. Examples for water utilities are the conflicting mandates for achieving both environmental quality objectives and effective price regulation, and improved levels of wastewater treatment at the same time that the system is to be expanded.

> Box 2. **Possible structure for a Water Sector Reform Group (WSRG) at national level**
>
> *Objective* – to prepare detailed proposals for the reorganisation of the sector and then to co-ordinate the implementation of the approved policy.
>
> *Pre-requisite* – policy decision by government to restructure the water sector in order to meet long-term objectives.
>
> *Established by* – the Minister responsible for the sector.
>
> *Reporting to* – the group of ministers with a major influence on the sector; in most countries these are the 'water' minister, the minister of environment, the finance minister, and the economics minister. They would meet as required to make important decisions, but otherwise they would meet monthly to review progress and discuss interim conclusions etc.
>
> *Members* – at a management level, there would be an inter-ministerial working group of senior officials, representing the relevant ministries and agencies in the sector. Working parties to be established reporting to the WSRG to consider specific detailed issues, such as the options for consolidating the water utilities.
>
> *Milestones* –
>
> 1. *Preparation of a water reform policy paper* – this will set out the detailed reasons why reform is necessary, what is involved in the reform programme, what changes in legislation are needed, the pros and cons of implementing the programme, and a broad action plan together with a timetable.
>
> 2. *Promotion of the net benefits of the reform programme to all stakeholders.* This includes promotion of the benefits to other ministers, senior officials in the departments affected, the municipalities and management of the water utilities as well as customer organisations.
>
> 3. Approval of the reform programme by the government and the legislature, as appropriate.
>
> 4. Other milestones as determined by the detailed action programme.
>
> ---
>
> Source: *Obstacles and Opportunities to Commercialising Urban Water Services in the New Independent States (NIS): Final Report* by Environmental Resources Management (ERM) under contract to the UK Department for International Development (June 2000).

World-wide experience with community development programs suggests that decentralising responsibility for infrastructure to subnational levels of government and actively involving local beneficiaries can improve infrastructure performance. Local monitoring can be effective for ensuring that officials perform diligently, and sanctions imposed locally against defaulting or free-riding community members can be more difficult to ignore. Decentralisation also makes it easier for people to obtain information on budgets and the use of funds, thus increasing the transparency of public actions and reducing corruption.

Comprehensive water sector reform should better define the roles of various government institutions, and provide vodokanals with more autonomy, assuming appropriate frameworks are in place for transparency and accountability of their operations. The government's role would essentially shift from that of service provider to overall regulator of the sector with responsibility for:

- Creating the economic, financial, and legal conditions in which utility management can meet the long-term objectives of the sector.

- Monitoring the performance of the utilities and to take corrective action within the agreed regulatory framework to address shortcomings in performance.

To avoid duplication between the different levels of government, experience in other countries suggests that:

- The *central government* should set the national policies and implementation measures through which others are expected to act by their own interests and incentives to achieve the long-term objectives for the sector.
- Each *local government* should focus on the organisation of the water sector in its area of constituency.

The following sections attempt to further explain the types of institutional problems encountered and how these may be addressed, by suggesting appropriate roles for the national authorities, the local authorities, and the vodokanals.

3.1. National Authorities

In many NIS, provision of urban water supply and sanitation services falls within an institutional/regulatory framework that is largely the responsibility of the national authorities. Clarification of roles among the many government bodies, and between the central authorities and the direct service provider, the vodokanal, will promote efficiency and accountability in the sector.

In some cases, problems with the current system can be alleviated by further decentralisation to place government responsibility for service provision directly with local governments. However, care must be taken to avoid excessive fragmentation and dispersion of resources: there may be instances where management may be more efficient at a regional level, due to particular characteristics and economies of scale in a region. In general, though, placing the authority with local governments increases the potential for responsiveness to local constituents, and allows national or sub-national authorities to assume, and focus on, regulatory oversight and sector development. Also, if the local government is to assume certain financial responsibility for the sector, this must be accompanied by the necessary shift in fiscal authority from the central to the local level, so that municipalities have legal authority to raise necessary revenues.

The recommended responsibilities to be assumed by the national authorities are outlined in Box 3.

Rather than placing key responsibilities with multiple government institutions, the central government's regulatory oversight roles may best be served by establishing an independent regulatory agency, similar to those that exist in other countries. One advantage of this approach is that an independent agency is less likely to be subject to political interference. However, for an agency to do its job effectively, there is a need to ensure that the regulator has sufficient access to operations information from water utilities, such as detailed cost and operations data. This goal may be addressed by building information disclosure clauses into water utility performance agreements.

The staffing of any regulatory agency will be an important consideration. Diverse technical expertise will be required, including economic, financial, and legal skills, in addition to the traditional engineering skills that are commonly found among current staff of national water departments.

In addition to establishing a solid institutional and regulatory framework, the central government also has a responsibility to ensure that a legal framework is in place to clarify and support various stakeholder roles, responsibilities, and interactions. For example, the legal status of vodokanals and their relationship to local governments is often unclear, potentially limiting the utility's ability to enter into service contracts with outside parties. The ambiguity also extends to issues such as property rights for infrastructure assets. In cases where national governments still retain ownership of all assets, the rights to manage and use the assets is not always clearly established by regulation. With such rights undefined, it is difficult for either the municipality or the water utility to use these assets to obtain financing necessary for investments. This provides a barrier to effective management of the water utility and increases the level of operational dependency between the water utility and the state.

Another aspect of water management that is best managed at the central or regional government level is the allocation of water resources among different competitive uses (domestic, industrial, agricultural), by providing principles and rules for resource management. Several countries throughout the world have adopted a "river basin management" approach to management and planning, which

> **Box 3. Recommendations water sector responsibilities for national authorities**
>
> Decentralisation:
> - Decentralising responsibility for water supply and sanitation services to the municipal level, avoiding excessive fragmentation.
> - Establishing the legal, regulatory and institutional framework for sound and sustainable municipal finance, including effective planning, supervision and fiscal control systems for municipalities.
> - Clarifying the legal status of vodokanals, their relations with local governments and property rights for infrastructure.
> - Establishing a framework for treating the inherited debts of vodokanals.
>
> Regulatory Oversight:
> - Depending on the particular circumstances in a country, consider establishing an independent, national regulatory agency to ensure that vodokanals do not exploit a monopoly position and/or to protect them from undue political interference. In such cases, the objectives of the regulation should be clearly identified and appropriate means for achieving them provided.
> - Regulating issues that have national or inter-municipal dimension, such as standards for environmental quality, wastewater discharge and drinking water; and establishing the legal framework to facilitate water and sanitation management initiatives undertaken jointly by groups of municipalities.
> - Establishing the legal and regulatory framework for stakeholder involvement, including private sector participation and consumer protection.
> - Establishing a framework for managing the competitive uses of water at the national and regional levels, including principles and rules for the management of different water resources, and policies for integrating municipal water and sanitation systems into coherent programs for water resources management within river basins.
> - Ensuring that an adequate system for monitoring water quality is in place and that the results are available to the public.
> - Strategy Formulation and Technical Assistance
> - Defining strategic policies and development objectives, including investment strategies and the means for financing them; such policies and investment strategies should strike an appropriate balance between water supply and sanitation objectives.
> - Providing assistance to utilities and local governments in areas such as capacity building, finance, and international assistance co-ordination.
> - Promoting demonstration projects to reform selected vodokanals; disseminating results; publishing performance indicators for vodokanals.
> - Facilitating market creation and promoting competition in the supply of goods and services to vodokanals.
>
> Source: *Obstacles and Opportunities to Commercialising Urban Water Services in the New Independent States (NIS): Final Report* by Environmental Resources Management (ERM) under contract to the UK Department for International Development (June 2000).

integrates municipal water and sanitation systems into coherent programs for water resources management based on river basins. This is a model that the NIS will want to consider when setting long-term sector objectives, as it could place some limitations on the scope of urban water management decisions taken at the municipal level.

The central government could also be key to providing demonstration projects; creating competitive markets for goods and services supplied to water utilities; and establishing a program to manage debts which many vodokanals have inherited from the previous system. (This is further discussed in Section 4.)

3.2. Vodokanals and Local Authorities

Under the current situation, the state often retains full or majority ownership of the utility, even though it also has regulatory, social, and financial responsibilities. Besides limiting the utility's autonomy, this situation limits the state's ability to maintain an "arms length" regulatory relationship with operators. As a result, many management issues are politicised. Operational decentralisation for water services should be accompanied by fiscal decentralisation, so that local government also has authority over water sector related finances.

Management of the vodokanals should be the responsibility of autonomous legal entities to allow for more flexibility and accountability in managing the sector, and to minimise political interference. There is a need to make explicit the legal relationship between the vodokanal and the relevant government authority, most often the municipality. This includes full identification of the cash flows between the two parties associated with the delivery or urban water services, an arrangement often termed "commercialisation". This step can be a precursor to wider change in the structure or ownership of the utility, resulting in increased financial and operational autonomy.

There should also be efforts to better define performance agreements between the vodokanal and government, as these are key documents guiding water utility management and are vehicles for specifying the roles of each party in the short-term. One example is the vodokanal "charter", a legacy of the Soviet era. This document is usually ambiguous with regard to financial and technical performance targets, and does not address terms and conditions on which the utility's performance can be evaluated, such as price, quantity, and quality. New arrangements need to be specified between the government and autonomous water utilities. Well developed agreements can protect the public from abuse by the utility, and can also include adequate restraints to protect the utility from discretionary actions to by the regulator. Discretion can be reduced, for example, by introducing into the provider's contract negotiated tariff adjustment formulas, agreed actions in relation to extraordinary events, a defined procedure for settling disputes, etc. An agreement could also specify who is responsible for making investments and for providing financial support to low-income households.

Another form of utility dependency is operational dependency, where the utility must rely on a state-owned or state-controlled third part organisation for bulk water supply. While this situation may persist even with an independent utility, there should be clearly defined pricing and supply agreements (subject to regulatory oversight), which allow the water utility to base their management decisions on predictable water supply conditions.

Consumer co-operatives in some Western European countries have proven to be an efficient way of organising supplies of water, electricity, and district heating, notably for smaller towns and communities. They leave local communities with full powers of setting supply standards, establishing the economics and logistics of water provision, freedom of tariff-setting and collection. This approach has generated position experiences in carrying out otherwise unaffordable investments. They have also proven to promote cost-awareness and efficiency, improve willingness-to-pay and acceptance of otherwise unacceptable enforcement measures, and led to more responsible consumer behaviour. Such co-operatives can make consumers take full responsibility for their own service demands – while also reducing political and financial pressures on municipalities and central government.

3.3. Public Input

Throughout the region, vodokanals lack the experience and institutional capacity necessary for interacting with consumers in a market-based setting. For example, public consultation (in the form of affordability and willingness-to-pay surveys), is rarely incorporated within the actual tariff setting process, and customer complaint procedures can often involve a confusing array of organisations, none of which considers itself ultimately accountable for addressing the complaint. To better serve the public, there is a need to build consumer service skills, and to rationalise and simplify complaint procedures where possible, including establishing a focal point for receiving complaints against the industry.

Willingness to pay (WTP) surveys find that people are generally willing to pay more for an improved water and wastewater service in NIS cities (see Table 1). However, the size of willingness to pay responses varied immensely between locations.

Table 1. **Some WTP results from the NIS city affordability studies examined**

Location	Date of survey	Average WTP for an improvement in water supply
Baku	1994	10 times more than current bill
Tbilisi	1998	0.5 times more than current bill
Chisinau	1995	twice the current bill
Kaliningrad	1997	47% would be WTP a 20% rise in the bill
Several studies, Ukraine	1996, 1997, 1999	twice the current bill for a 24hr supply

Source: *Obstacles and Opportunities to Commercialising Urban Water Services in the New Independent States (NIS): Final Report* by Environmental Resources Management (ERM) under contract to the UK Department for International Development (June 2000).

There is need to maintain a dialogue with the public throughout the reform process and to seek their input as key service decisions are made. With regard to financing, the government at all levels needs to clearly communicate to the population the sector's new financing and subsidy policies. It is easier to increase and collect tariffs, if people fully understand and believe that the government no longer has the resources to support the sector, and that the future quality of service depends on consumer willingness to take responsibility for their systems, including paying for services received.

Finally, local governments have an opportunity to develop constructive partnerships with public and consumer groups. Such groups can play an active role in educating consumers about the importance of water saving and avoiding pollution. Non-governmental organisations can also monitor the behaviour of polluters and provide substantive inputs to policy development and decision-making.

Box 4. **Public participation and protection of the poor and consumers**

Reform of the urban water sector will change the role of the public from essentially passive consumers of state-provided services, to purchasers of these services. This will require a greater involvement of the public in decisions concerning the level and type of services provision, and the associated financial implications. Government and vodokanals should provide for effective public involvement in decision-making in sector reform and ensure that poor and vulnerable groups have adequate access to water services.

- National authorities should establish a legislative basis for public participation in key decisions concerning water supply and sanitation services.
- Local government and vodokanals should actively provide consumers with full information and opportunities to participate in key decisions concerning water supply and sanitation, especially through public meetings and participation in decision-making bodies.
- Performance contracts between local governments and vodokanals should be developed through a participatory process and ensure that the interests of all stakeholders, particularly the poorest and most vulnerable, are protected.
- Government, and not vodokanals, should have the ultimate responsibility for ensuring that poor and vulnerable households have adequate access to water and sanitation services; transparent, targeted and efficient subsidies which take account of tariffs for all utilities and address integrated household needs should be used to provide support to such households.

3.4. *Private Sector Partnerships*

Many opportunities for improved operating efficiency and investment financing are available to water utilities that enter into partnerships with private sector enterprises, as has occurred in other parts of the world. An important obstacle for attracting private sector finance is that most NIS countries do not have a transparent business environment and supporting legal framework in place. This often presents too much risk for the potential investor to invest in NIS water and wastewater utilities. While the decision whether or not to pursue private sector partners is ultimately up to local communities, it will only be possible if the national authorities have the appropriate framework in place, and have the necessary regulatory capacity to enforce and monitor contractual agreements. There are several different types of partnership arrangements, which place different degrees of financial and operational responsibility with the private sector partner. Case study experience from implementation of the various models yields some important lessons for how to make them work most effectively. Municipalities should also retain the assistance of expert legal, technical, and financial advisors with appropriate experience. Annex I contains recommendations for successful involvement of the private sector in the provision of water supply and sanitation services, and the background paper *Private Sector Participation in Urban Water Supply and Wastewater Financing and Management: An Opportunity for Increased Financing and Improved Efficiency* (CCNM/ENV/EAP/MIN(2000)7) provides a more in-depth discussion.

4. Financial Management of Vodokanals

Most vodokanals were not operating in a financially sustainable way prior to 1991, and the situation has continued to deteriorate during the reform period. Their financial resources will not allow them to maintain current levels of service into the future. The following subsections outline some of the key obstacles that need to be addressed in order to improve the financial status of the water sector and to allow it to operate on a financially sustainable basis over the long-term.

4.1. Public Good versus Economic Good

A major psychological obstacle to water sector reform is that water is still viewed and treated as a "public good" rather than an "economic good" in most NIS. This belief implies that water should be available to and consumed by all citizens without limitation. Consequently, current water and wastewater pricing policies fail to appropriately value water and related services as a limited resource. Moreover, the view of water as a public good often means that political and social considerations take precedence with decision makers on issues such as tariff levels. While social considerations should certainly be incorporated into water sector reforms, programs to help low-income consumers pay for water should be managed by an appropriate government social agency and separated from the financial operations of the water utility.

The fact that the general public does not view water as an economic good contributes to problems of water wastage in the system. The lack of individual meters as a basis for household charges in apartment buildings, means that customer tariffs are not based on actual consumption amounts, but often on a fixed tariff amount, which may be calculated using the number of people per household and formulas of consumption norms. This creates no incentive for consumers to conserve water. A new volume-based pricing strategy should be developed which considers the cost-effectiveness of various metering options. The cost of investing in individual meters can be weighed against their potential benefit in reducing water consumption. In the case of large apartment complexes, it may be that the most technically, financially, and administratively viable option is block metering. For the short-term, an intermediate option, such as installing meters only for larger customers and those know to waste large volumes of water, can also be considered.

In addition to demand management techniques, such as tariff reform, awareness raising campaigns should be designed to encourage more efficient use of water by consumers.

4.2. Tariff Collection

In general, low collection of water and waste water tariffs, and payment in non-cash forms such as barter and off-set,[2] is a critical and complex issue facing the water and sanitation sector. A 1998 study found that the majority of vodokanals in the Russian Federation collect less than 40% of their billings, and some less than 20 per cent.[3] Some of the worse offenders are state-owned enterprises, such as government departments and the defence industry, yet the utilities are not legally permitted to disconnect budget organisations or major industries. The ability of the sector to achieve long-range financial stability is directly related to this issue, as well as the simple ability to have enough cash on hand to pay basic expenses such as salaries. It has trapped many vodokanals in a web of every increasing arrears and receivables, compounding their other difficulties in obtaining solid financial status.

The move away from barter and off-set systems is linked to overall economic reforms and will require a concerted effort between the government and the vodokanals. At the same time, the coverage and efficiency of billing and collections should be progressively improved. Presently, water and wastewater tariffs are often collected as part of a wider service charge bill handled by a separate agency. Collection efforts will undoubtedly be improved where the vodokanal is directly responsible for billing and collection, even if it may then choose to employ outside billing and collection agencies, and where the vodokanal has authority to impose sanctions for non-payment, and to cut-off service as a last resort.

4.3. Tariffs

The current approaches to tariff setting employed in the NIS do not always promote economic efficiency. A "cost plus profit" methodology is often used to estimate the total costs for water supply for a set period, and consequently what the water supplier may charge to customers. The allowed profit is set anywhere between 10 and 40% or more, and then often taxed. This approach creates no incentive for the supplier to reduce operating costs, as it only reduces their profits. Alternative pricing formulas offer greater incentives for improved operating efficiency and should be considered during reform. For example, a formula where the utility is entitled to keep any additional cost savings from improved operational efficiency beyond an agreed upon target. Thus, the greater the reduction in production costs, the more profits are maintained by the water utility, without increasing the agreed upon total cost of service for a given time period. Such a formula could, for example, be incorporated into a performance agreement between the municipality and the utility for a period of 3 to 5 years. This arrangement is a form of the "price cap" approach, whereby tariffs are maintained at a certain level, while profits are allowed to vary based on efficiency. Other formulas include a "rate of return" approach, where an operator's maximum allowed profit is set by the regulator to provide a fixed return on investment," and a "sliding scale regulation" where unexpected profits or losses are shared between the operator and its customers.

To determine what water supply and wastewater tariff rates are charged to individual customers, an "average cost" pricing model is often used in the NIS. Under this approach, the total anticipated cost of supplying water for the forthcoming year, (including both the actual costs and a profit for the supplier as explained in the preceding paragraph), is divided by the projected output of water to be supplied, to arrive at a per unit tariff. This means that it is difficult for the company to charge an economically efficient or financially viable price per cubic meter, because the allowable operational costs do not reflect all long run costs of supply.

Another pressure on cash flow is that the vodokanal is often subject to a variety of taxes, on both their actual or projected gross revenues and on their profits. The taxes, which can range from about 20 to 40% of the allowed profit margin, must be paid in cash and are often required in advance of revenue being received from customers.

Finally, high levels of inflation over the past 10 years or so have proved problematic, worsening the cash situation for vodokanals, as tariff adjustments have been typically sluggish and ex post. There is no standard method for adjusting charges, and changes to the tariff can take a long time to process and approve. An agreed procedure for periodic tariff revision and adjustment, indexed to inflation, should be established.

4.4. Subsidies

Financial efficiency can be enhanced by reconsidering the system of subsidies, through which the utility uses tariff revenue from one customer category to subsidise service to another category. This occurs because the operator may be legally required to discount water and wastewater bills for certain customer groups, due to a wide series of social objectives. For example, in the current situation, lower tariff rates for domestic customers may be cross subsidised by relatively higher tariffs for industry, often up to 10 times more. This practice can stifle the competitiveness of industry and result in a situation where many industries simply do not pay their bills. Often, the state does not fully compensate the water operator in cash for any losses, and this places a financial burden on the utility.

There is a need to redesign the current subsidy system, considering alternative ways to ensure protection of low-income customers. The redesign should involve multiple stakeholders, and an income support should be administered through a transparent process that is the responsibility of a social agency within the government, rather than the vodokanals. For example, income support can be channelled to low-income customers through the use of "water stamps" provided by the government to means-tested customers, which can then be used to pay part of their water bills. This type of approach makes the subsidies visible and transparent, and removes the responsibility for social welfare from the utility mangers, who can instead focus on running the company efficiently.

4.5. Accountancy and Investment Management

The physical condition of urban infrastructure networks varies from quite bad to critical. Inadequate depreciation funds and maintenance funds are unable to sustain the current networks. In Russia, for example, maintenance funds have been depleted over time because they have been used to fund operations as a supplement to inadequate tariff revenue, following municipal decisions to reduce tariff levels. Hence, most vodokanals have depleted or inadequate maintenance funds and a limited ability to ensure their gradual development again over time. This is related to the more general issue of the principles of accountancy followed in the NIS, which militate against adequate provision for depreciation. There is an urgent need to introduce international accounting standards which, *inter alia* would demonstrate the bankruptcy of most vodokanals in the NIS, something concealed by the accountancy principles currently used in the NIS.

The ability to build up depreciation and maintenance funds is directly linked to tariff revenue, and subsequently to tariff levels. In many cases, the government (at the local or central level) decides what fixed or variable costs can be included in the water/wastewater tariff level calculation, and these formulas frequently undervalue costs. Often, "bad debts" are not included as a cost for purposes of the water/wastewater tariff calculation, and allowances for depreciation of existing assets are too low to relative to the current asset replacement cost. What are included as costs, however, are several mandatory payments and taxes to the state, which can be numerous.

Vodokanals frequently have little or no say in the prioritisation and planning of capital investments (the preserve of state or municipal government) or in their technical specification and design (the preserve of state-owned design institutes). Plus, the allowed profit margins are often too small to finance anything other than minor investments or repairs. It is also difficult, and often illegal, for the operator to use profits as a means to acquire debt finance investments in the water supply network.

With limited funds available, there is a need to prioritise future investment and funding for interventions, starting with a careful assessment of the present system and its functioning. The background paper on *"Financing Strategies for the Urban Water Sector in the NIS"* (CCNM/ENV/EAP/MIN(2000)4) provides a methodology that can help guide investment decisions in the water sector, as well as other areas of environmental management.

4.6. Financial Management

At present, vodokanals have little or no financial autonomy and limited input to tariff setting or financial investment strategies, as the state usually has the final responsibility for tariff-setting and approval. Ideally, the vodokanal would have more direct authority to set tariff levels in accordance with management objectives, under the oversight of a state managed regulatory body, and according to an agreed tariff formula between the water utility and the state. At the same time, vodokanals should be subject to hard budget constraints, and any financial support to be received from the local government should be specified in the performance contract between the two parties.

To summarise the above discussion, a comprehensive system of improved financial management should focus on:

- Increasing public awareness of the economic value of water and encouraging water conservation.
- Adjusting tariff levels to eventually achieve full cost-recovery of operation, maintenance, and investment costs and to provide the operator with financial incentives to improve efficiency.
- Phasing out non-cash forms of payment and increasing overall payment collection.
- Designing financial support programs to address affordability constraints of low-income households and revising the current subsidy system to shift responsibility for income support to the government.
- Reducing operating expenses by seeking improved efficiency and reduced waste in all aspects of management.

A particularly difficult problem, that can create a major obstacle for vodokanals to move towards financial viability, is the high level of debt that many vodokanals inherited from the Soviet era. A comprehensive and targeted plan should be designed to deal with these debts, involving the water utilities in co-operation with the central government. Options for long-term debt management include debt reorganisation and, ultimately, debt relief from the central government. In the case of short-term accounts receivable, various forms of one-time state subsidies may be considered.

4.7. Private Sector Finance

There are several ways to attract private sector finance, including an offer of equity stakes, or potential arrangements involving a strategic investment partner, most likely an international water company. (See the background paper on "*Private Sector Participation in Urban Water Supply and Wastewater Financing and Management: An Opportunity for Increased Financing and Improved Efficiency*" (CCNM/ENV/EAP/MIN(2000)7) for more information. There are at the moment no examples of NIS water utilities that have successfully and unilaterally attracted private sector finance, although some are considering it.

5. Improving Efficiency of Vodokanals

The sustainability of the water sector in the NIS depends on the ability to improve system efficiency and to use resources in the most cost-effective manner. In addition to addressing the institutional and financial issues discussed above, which will help to establish the necessary framework conditions for improved efficiency of the sector, other operational reforms are also needed. To begin with, optimisation of the service area covered by each utility should be a priority. The optimal boundaries of service areas need to be based on an assessment of management capacity in the country or region, existing water system infrastructure, existing transport infrastructure, concentrations of urban areas and other factors. Based on this, service areas should be proposed which make optimal use of transport routes (for maintenance activities), current infrastructure (main pipelines pumping stations, treatment plants, etc.) and management capacity, thereby saving costs where possible. This may require the rationalisation of some water utilities, merging their management structures while maintaining existing infrastructure.

5.1. Standards

Service standards, including environmental and health standards, are an important factor in determining the costs, and ultimately the cost-effectiveness, of water and wastewater services. The current very strict standards are generally unaffordable and unachievable with current resources. While these strict standards can be maintained as long-term targets, but more appropriate levels should be adopted as realistic goals for the short-term. National standards which are stricter and more costly than those established internationally should be reviewed and modified to reduce compliance costs to a more realistic level. This change will need to be carefully negotiated, as it may meet with political opposition.

Health-based international standards for water quality can provide a good starting point. Over the long-term, techniques such as cost-benefit analysis can be employed to evaluate different water quality standard levels. For example, what additional benefit will be realised in terms of public health by raising water quality standards to a higher level, and how can this be translated into a monetary figure to be compared with the projected costs of achieving the higher standard? Ultimately, this type of analysis may conclude that some standards should be set at a lower level than those in more developed countries, based on what can be achieved with the current financial resources of the sector. More realistic standards will also increase the legitimacy of such standards in performance agreements and make it easier to hold water utilities responsible for achieving them.

Beyond the question of standards, reducing and optimising service delivery costs should be a structural priority in the short to medium-term. Some low-cost and easy to implement actions may be taken in the short term to reduce operating costs, but other cost and time intensive measures will also need to be taken. Specific attention should be paid to reducing leakages in the distribution system, which require an inflated volume of water to be supplied to meet a given level of customer consumption.

5.2. Staffing and Management

One of the most important issues facing the water sector is the need to develop human resource capacity to develop policies and practices based on modern utility management systems. There is an urgent need to diversify the skill base of the management team to improve efficiency on all levels. One way is to keep the core of the existing management team, but to also introduce new people with varied skills, and to provide training opportunities for all staff. A restructuring and reduction of staff may be necessary to achieve efficient staffing levels.

In order to broaden the skills base that exists among most vodokanals staff, capacity building is recommended in the areas of business management, economics, law, consumer awareness. Staff skilled in more specific techniques, such as conducting needs assessments and facilitating co-operation with the public, are also needed to implement many of the needed reforms in the sector. In addition, to expand on the technical expertise that exist in the water utility, the revision of engineering curricula of universities also requires attention, so that new staff come in with the latest technical training.

The management staff of vodokanals need be given incentives to achieve the highest possible levels of operational efficiency and to keep costs to a minimum. Suitable incentives can be built into performance agreements, as noted above. Government should then defer from interfering in day-to-day matters of the utility, as this removes responsibility for decisions from the managers. It is then very difficult to measure the performance of the utility managers and to know how to reward them.

In many countries, accounting is still seen by water companies as a tax-driven exercise rather than as a management tool. In the current system, there are several distortions that allow water companies to be presented as profitable enterprises, including the failure to fully account for bad debts or doubtful receivables. These type of accounting systems fail to present the real financial picture, and therefore impede necessary financial reforms. A transition towards use of international accounting standards is needed, both to better inform internal management decisions, and also to assist the utility in entering into contractual arrangements with outside partners for management and financial services. Use of international systems would allow for the independent audits of the utility. A move from manual record keeping to computerised information systems is also necessary to track cash flows on a more timely basis.

6. Sequencing Reforms

Reform strategies need to be integrated with, and take full account of, the broader process of economic, political and social reform in the NIS. Additionally, efforts to reform the water supply and sanitation sector should sequence actions and prioritise those on which other steps depend. Since some changes will require legislative or other time-consuming processes, they should be initiated as soon as possible, or accelerated if already in progress.

The temptation to rush towards immediate new legislation should be avoided. Often, new legislation is viewed as a means in itself, rather than as an instrument to support collective decision-making and the implementation of agreed objectives. This creates situations where legislation is developed and adopted before serious discussions have been held as to what the legislation is supposed to achieve and what complementary instruments are needed to achieve the envisaged results. This can actually hurt, rather than help, reforms, as it is often argued that policy changes cannot be implemented because it conflicts with old, or even newly approved, legislation. An example is a piece of legislation that prohibits private sector participation in water utilities, which may have been adopted before serious discussions have been held about the disadvantages and advantages. Also, the frequency with which legislation can change within some countries creates confusion as to what legislation is current, and what has been repealed. This can significantly hinder the process of planning changes in management structures and ownership.

A program of reform should include the following priority actions:

- Launching participatory, multi-stakeholder processes to support the development and implementation of the strategies.
- Decentralising authority to the local level.
- Creating autonomous vodokanals and, as required, an independent regulatory agency.
- Establishing performance contracts between local governments and vodokanals, initially with a focus on improving service levels through affordable, low-cost measures.
- Strengthening the financial stability of vodokanals by reforming tariff levels, collection systems and debt.

Notes

1. This paper draws heavily on recently completed reports: *Azerbaijan Water Supply and Sanitation Sector Review and Strategy* by the World Bank Infrastructure Sector Unit, Europe and Central Asia Region and *Obstacles and Opportunities to Commercialising Urban Water Services in the New Independent States* (NIS): *Final Report* by Environmental Resources Management (ERM) under contract to the UK Department for International Development (June 2000).
2. An off-set occurs when accounts payable between two entities are reduced by equal amounts, but no actual payments is made bertween the two.
3. Parker, D., "Water and Waste Water Services in the Russian Federation: A Study of Four Vodokanaly," *Post-Communist Economies*, Vol. 11, No. 2, 1999, pp. 219-235.

Appendix 1

Recommendations for Successful Involvement of the Private Sector in the Provision of Water Supply and Sanitation Services

1. **Define Clearly the Objectives:** The Government should develop a clear vision of the strategic objectives and their relative importance. These objectives are likely to include improved operational efficiencies and cost reduction, service modernization, rehabilitation of the existing infrastructure, and mobilization of private capital for new investments. The early definition of the strategic objectives will facilitate the selection of the most appropriate instrument for involvement of the private sector and the preparation of the bidding process.

2. **Build Internal Consensus on a Specific Option:** Description of the options and strategic objectives to build consensus is a key step that should not be overlooked in the rush to get into details of the contract and reaction to unsolicited proposals, which eventually results in delays, frustration and dissatisfaction at later stages. All stockholders should be involved in this process (local political leaders and administrators, vodokanals, employees, customers).

3. **Adopt Realistic Targets for Service Levels, Standards, Investments and Tariffs:** Adopting affordable service objectives and standards and describing them in the contract and bidding documents is essential for obtaining responsive, comparable bids. All investments should be rigorously evaluated and its implications on the tariffs and financial viability of the water and wastewater enterprise should be studied carefully.

4. **Deal with Labor Issues Early On:** The employees may be apprehensive of the involvement of the private sector. However, experience has shown that the operators tend to retain a large proportion of the existing workers and wages usually go up. Even so, the very purpose of private participation is to enhance productivity, which often will mean doing a better job with fewer people. Many tasks currently performed by the vodokanals should be outsourced. Even without the involvement of a private utility operator, the Government will sooner or later be confronted with tough labor decisions and will have to balance the interests of the workers with those of the entire population demanding better services. As part of the process to involve the private sector in the provision of water and wastewater services, social safety plans should be prepared.

5. **Ensure that the Private Operator has the Freedom to Perform Well and Appropriate Regulation is in Place:** The Government should understand the level of freedom it is necessary for the private operator to deliver its obligations. Also, it must create and legally allow the regulatory structure to supervise the operator in order to protect the public interest, create incentives for operational efficiency.

6. **Use a Transparent, Competitive Process to Select the Private Utility Operator:** Competition is crucial to get the best proposal the market can offer and to increase the credibility of the process. Private firms often prefer negotiated deals without competition. Competitively procured contracts are likely to result in lower costs, better qualified operators and transparency. International experience has shown that competitive bidding can be as quick as direct contracting.

7. **Retain Expert Independent Advice:** Contracting a private utility operator is a difficult task requiring very specific experience, which is currently not extensively available in most NIS. Therefore, it is necessary that interested Governments retain the assistance of experienced legal, technical and financial advisors of the highest international quality for the preparation of the contract and bidding documents and for the full duration of the contracting process. Their tasks should include: investment program, financial analysis to determine cost and tariff implications, definition of regulatory arrangements, employee transfer or compensation arrangements, assistance in the bidding process, and continuing advice through negotiations and closure with the selected operator.

Chapter 5

PRIVATE SECTOR PARTICIPATION IN URBAN WATER SUPPLY AND WASTEWATER FINANCING AND MANAGEMENT: AN OPPORTUNITY FOR INCREASED FINANCING AND IMPROVED EFFICIENCY[*]

1. Introduction

The urban water sector presents difficult economic and political choices for governments. Traditionally this engineering-dominated sector is plagued by a long history of under-pricing, and opposition to pricing for "moral" and social reasons. These factors have contributed to the unwillingness of many governments to acknowledge water as a finite natural resource and an economic good – a commodity that needs a market price reflecting its true value to society. This often results in:

- Inefficiency: One-half of the drinking water that enters the system is lost or otherwise unaccounted for before reaching its customers in many NIS.

For instance, in Tbilisi, Georgia, it is estimated that up to 700 km of pipeline has seriously deteriorated, and as little as one quarter of supply actually reaches the consumer. In Zaporizhzhia, Ukraine, 25% losses are recorded, but as metering is not widespread, real losses are estimated to be closer to 50%. Finally, in Dushanbe, the capital city of Tajikistan, losses are estimated to be as high as 70%.

- Unreliable services, lack of coverage, sporadic maintenance, poor design: this is mainly due to lack of managerial accountability, hard-budget constraints, and the absence of commercial practices in many public infrastructure agencies.

For example, in Baku, Azerbaijan, the use of steel pipes has created a significant public health hazard. In Kazakhstan, where part of the water supply network in the Aralsk region (constructed in 1980) is above the frost line, interruptions of supply due to cracked or frozen pipes are frequent. More generally, throughout the NIS, inadequate water supply in cities is often attributable to poor maintenance of pumps (or lack of funds to pay for the electricity to run them), particularly in high rise buildings served by booster pumps, where residents above the second or third floor often do not receive any water.

Improving the delivery of urban water and waste water services is a critical need for many developing countries and economies in transition, requiring greater efficiency through better management and significantly increased investments. However, this is increasingly difficult as both government budgets and ODA have shown decreasing trends recently. Some governments are therefore increasingly looking to a range of private sector partners to provide access to two key resources: 1) improved management systems and technical options, and 2) private investment funds.

---

[*] This paper summarises a report that was prepared for the OECD by Bradford Gentry and Alethea T. Abuyuan from Yale University (OECD, Global Trends in Urban Water Supply and Waste Water Financing and Management: Changing Roles for the Public and Private Sectors, CCNM/ENV(2000)36). The present paper therefore draws extensively on that paper, including some of the wording. For further details and references, we refer the reader to the original paper.

© OECD 2001

1.1. Declining government budgets

One reason for changes from public to private sources in infrastructure finance is that many governments are finding the burden of public finance increasingly difficult to bear. It is estimated that developing countries for example, spend about US$250 billion a year on new and rehabilitated infrastructure; 90% of this amount is taken from government tax revenues or intermediated by governments through foreign financing (both concessional and non-concessional funds from multilateral and bilateral sources). About 30% of the amount (US$65 billion) is spent each year on water sector infrastructure such as: hydropower (US$15 billion); water and sanitation (US$25 billion); and irrigation and drainage (US$25 billion). The situation is similar in the NIS, where the state has traditionally been the exclusive source for finance. In Georgia alone, where approximately 80% of a population of 5.4 million are connected to centralised water supply systems, annual expenditures (including investment and operational costs) of about US$178 million are needed to simply maintain current (unsatisfactory) levels of service. If these figures were extrapolated to the overall NIS population of 280 million about US$10 billion would need to be spent annually on water supply and sanitation to maintain current levels of service.

In addition, governments find their tax revenues insufficient to meet all competing needs – such as social welfare programs, health systems and defence. NIS are faced with the particular situation, where budgets have continuously contracted since the beginning of the transition process, due in large part to the collapse of output and GDP. In 2000, the real GDP in the NIS is only 50% of its level at the beginning of the nineties. This explains why public investment in urban water systems in the NIS have been close to zero during the last decade.

In this NIS, the shortage in the availability of government resources for investment, particularly in the water sector, is compounded by opposition to the implementation of the polluter- and user-pays principle, and growing industrial activities for access to a finite water resource base; the lack of political will to change existing allocation patterns in the face of increasing scarcity; and increasing pollution of water sources. The net effect of such factors has been at least a two-fold increase in the cost to government of raw water supply. At the same time, as mentioned earlier, much of the raw water is lost or constitutes otherwise unaccounted-for-water ("UfW") in many publicly operated water utilities.

1.2. Decreases in ODA

While government budgets have declined in many developing and transition economies, official development aid (ODA) has been stable or dropped in recent years. According to the World Bank, transfers of ODA to developing countries averaged around US$50 billion per year from 1990 through 1999, well below the US$125 billion target set at the 1992 Earth Summit. Recently published surveys by the OECD indicate that aid from developed to developing countries as a share of the wealth of the developed countries has also been limited. Estimates suggest that in 1998, while US$51 billion was provided in aid to developing countries, as a proportion of national income, aid from developed countries rose only slightly from 0.22% to 0.23%. This trend has been attributed to a number of factors including: *i)* other demands on the budgets of donor countries, frequently domestic in nature; *ii)* corruption in recipient countries; *iii)* lack of effectiveness of previous assistance to reduce poverty; and *iv)* poorly targeted aid projects.

1.3. Private capital flows and privatisation

In stark contrast to ODA, total global flows of private capital doubled in the first part of the 1990's, and private investment in developing countries increased six-fold – increasing from under US$50 billion in 1990 to over US$300 billion in 1997. Even with the recent global financial crisis, World Bank data indicates that net private flows to developing countries remained between four and five times larger than official flows in 1998 and 1999.

At the same time, fiscal constraints and increasing disenchantment with the performance of state-provided infrastructure has increasingly led governments to turn to private solutions for financing and providing urban services. Private markets have responded with vigour. According to a recent report by the World Bank (1999), cross border flows dominate infrastructure finance, even in countries with high national savings rates. This has been attributed to both the benefits investors gain from diversification, as well as to the underdevelopment of local capital markets in the countries concerned.

Total private financing of infrastructure in developing countries rose from less than US$1 billion in 1988 to more than US$27 billion in 1996. Much of this investment has been in telecommunications, power, and other large construction projects. Similarly, foreign direct investment (FDI) in urban water services has increased world-wide from US$297 million between 1984 and 1990 to US$25 billion between 1990 and 1997. However, FDI flows generally have been concentrated in a small number of countries; most have received little or no FDI. About 6% of FDI in the water sector has flowed to the CEE/NIS region, and it is likely that most of this money has actually been invested in the accession countries of the CEE rather than in the NIS. In the NIS only two projects with private sector involvement in the water sector are known as of today (one in St Petersburg, Russia, the other in Almaty, Kazakhstan).

1.4. Implications for water financing and management

In response to this new situation, many governments are exploring increased private investment in water and waste water services. By doing so, they are trying to expand their access to new financial resources, as well as to new technical and managerial skills.

Many different approaches are being tried. Some grow out of the experience in industrialised countries. Some are newly created to meet the needs of developing and transition economies. All require involvement by governments and users. The key is to find the right balance of roles to meet priority water needs in the local context. In particular, the concerns and fears that many people in the NIS have *vis-à-vis* private sector participation need to be addressed. These concerns include the possibility of monopoly abuse by the private water operator, the need to guarantee the reliability and quality of water services, the need to protect the most vulnerable sections of the population and the likelihood of staff reductions.

It is important to note at the outset that this paper is not meant to advocate private sector participation. Rather it is looking at the different options and drawing the lessons learned from the past experience. In doing so the paper points at many of the difficult choices that governments would need to make if they were to opt for some measure of private sector participation in the urban water sector, and presents their advantages and disadvantages. Section 2 of the paper describes a range of options of private sector participation. Section 3 distills lessons learned for attracting private investors and Section 4 offers suggestions for steps governments and others can take to increase private involvement in water and waste water services.

2. Developing countries and economies in transition: emerging models for private investment

A wide range of approaches for involving the private sector in improving the performance of water and sanitation systems exists. Some options keep the operations in public hands, but change the operational incentives. For example, in "corporatisation", the water utility remains in public ownership, but adopts a formal, corporate structure. Other options involve private actors in a variety of ways and to a variety of degrees, ranging from private operation only to private operation, investment and ownership. In all of these options however, the public authority remains responsible for overseeing the activity and for ultimately ensuring that public needs are met. Some of the major approaches being used are described below.

2.1. Private operation, public oversight, investment and ownership – service contracts

Under any of the various forms of service contracts (operations, management, sometimes leases), the government hires a private organisation to carry out one or more specified tasks. Such service contracts are often established for a period of five to seven years. The government remains the primary provider of the water service and only delegates portions of its operations. The private firm must perform the service for the agreed upon fee and meet specified performance standards. Governments generally use traditional competitive bidding procedures to award service contracts based on specific service requirements, which tend to work well given the limited time frame and narrowly defined nature of these contracts. For example, the operation of a water or waste water treatment plant, provision of water distribution services, meter reading, billing and collection operations, and the operation and maintenance of standpipes.

Financing Structures

Under a service contract, the government pays the private business a pre-determined fee for the service, which may be based on a one-time fee, unit cost, shared savings, shared revenues, or other formula. The private contractor does not typically have a business relationship with the end-users and all financial interactions are made directly with the government. The government is responsible for funding any capital investments needed to expand or improve the system. The only private capital invested is the contractor's fronting of the bid and other preparatory costs, which it will seek to recover through its fees.

Box 1. Strengths and weaknesses of service contracts

Potential Strengths
- Service contracts provide a relatively low-risk option for expanding the role of the private sector.
- Service contracts have great potential to provide improvements in performance and efficiency through technology transfer and the acquisition of technical and/or managerial capacity.
- Service contracts are one of the most competitive forms of private involvement, since contracts are reissued frequently. Also, because service contracts are limited in scope, the barriers to entry are fairly low.

Potential Weaknesses
- Service contracts do not involve significant infusions of private capital.
- Service contracts do not necessarily create a base from which to optimise entire water and waste systems.
- Service contracts leave the government in charge of many of the most explosive political issues (namely the fee and the expansion of network) and therefore do little to prevent from negative political intervention.
- Municipalities are often under pressure to award service contracts to the lowest bidder, with less consideration given to the businesses' ability to provide high quality service.

2.2. Private operation and investment, public oversight and ownership – greenfield "Build, Operate, Transfer" arrangements, concession contracts

If one of the government's key goals is attracting more private capital into the water system, but it is uncomfortable giving up ownership of water assets, two major techniques are available. One focuses on the construction and operation of new treatment plants – so-called "greenfield" or "Build, Operate, Transfer" ("BOT") arrangements and their variants. The other anticipates the construction of new facilities, but as part of the overall running of the entire water and waste water system, including customer billing and collection – so-called "concession" arrangements. Each is described below.

2.2.1. *Greenfield/BOT arrangements*

BOT contracts are designed to bring private capital into the construction of new treatment plants. Under a BOT, the private firm finances, builds and operates a new plant for a set period of time according to performance standards set by the government. The operations period is long enough to allow the private company to pay off the construction costs and realise a profit, typically 10 to 20 years. The government retains ownership of the infrastructure facilities and becomes both the customer and the regulator of the service. BOTs tend to work well for new facilities that require substantial financing. Governments generally issue BOT contracts for the construction of specific infrastructure facilities, such as bulk supply reservoirs and drinking water or waste water treatment plants. BOTs typically involve the construction and operation of only one facility and not the entire system.

- Financing Structures

Under BOTs, the private sector provides the capital to build the new facility. In return, the government agrees to purchase a minimum level of output over time, regardless of the demand. The purpose is to ensure that the private operator can recover its costs over the contract period. This requires that the government estimate demand with some accuracy at the time the contract is set. Otherwise, it will have to pay for water that is not being used, even if demand is less than expected.

Box 2. Strengths and weaknesses of BOT arrangements

Potential Strengths
- BOTs are an effective way to bring private money into the construction of new water and wastewater facilities or the substantial renovation of existing ones.
- BOT agreements tend to reduce market and credit risks for the private investors because the government is the only customer, reducing the risks associated with insufficient demand and ability to pay.
- The BOT model has been used to build new power plants in many developing countries, so there is a lot of experience with this contractual form already.

Potential Weaknesses
- BOTs generally involve only one facility, which limits the private firm's ability to help optimise system-wide resources or efficiencies.
- The duration and complexity of BOT arrangements make these contracts difficult to design; a fact that may eventually undermine the positive effects of the initial competition, when renegotiation is necessary to reflect changed circumstances once they are underway.

2.2.2. *Concession contracts*

In a concession contract, the government turns over full responsibility for the delivery of water and waste water services in a specified area to a private contractor ("concessionaire") – including all related construction, operation, maintenance, collection, and management activities. The concessionaire is responsible for any capital investments required to build, upgrade, or expand the system, and for financing those investments out of the tariffs paid by water users. The public sector is responsible for establishing performance standards and ensuring that the concessionaire meets them.

In essence, the public sector's role shifts from being the provider of the service to the regulator of its price and quantity. Such regulation is particularly critical in the water sector, given that water is a public good and piped delivery systems are natural monopolies. The fixed infrastructure assets are entrusted to the concessionaire for the duration of the contract, but they remain government property. Concessions are usually awarded for time periods of over 25 years. The duration depends on the contract requirements and the time needed for the private concessionaire to recover its costs and profit. Concessions have been used in many Latin American and CEE countries.

• Financing Structures

Over the life of the contract, the private sector manager is responsible for all capital and operating costs – including infrastructure, energy, raw materials, and repairs. In return, the private operator collects the tariff directly from the system users. The tariff level is typically established by the concession contract, which also includes provisions on how it may be changed over time.

Structuring the tariff and the accompanying regulatory system is often the most complicated part of any concession arrangement. Tariffs need to be high enough to allow the operator to make a profit if it performs well, but not so high that the profits are excessive – causing a political backlash. In tariff setting, the key battles are over information. Does the regulator have enough information to make informed judgements as to the actual state of the concessionaire's finances during the concession period? Is the private operator meeting the performance standards and are the customers well served? Managing information flow among the concessionaire, the users, and the regulators is one of the key challenges facing concession arrangements.

Box 3. Strengths and weaknesses of concession contracts

Potential Strengths

- Concessions are an effective way to bring private money into the construction of new water and waste water systems or the substantial renovation of existing systems.
- Combining the responsibility for investments and operations gives the concessionaire strong incentives to make efficient investment decisions and to develop innovative technological solutions.
- Concession operations are less prone to political interference than government-operated utility services because the service stays under the same operator regardless of changes in governments.

Potential Weaknesses

- Large-scale concessions can be politically controversial and difficult to organise, due to the need of consultation with interested parties.
- Government oversight of the concessionaire's performance against the contractual standards is critical. This often requires governments to expand significantly their regulatory capacity.
- It is difficult to set fixed bidding and contractual frameworks for concessions that are to last for 25 years or more.
- Some argue that the benefits of open competition are limited in the concession market since only a small number of large, international companies are usually able to meet the pre-qualification criteria for bidding to run a concession.

2.3. Public-private operation, investment and ownership, public oversight – joint ventures

In joint ventures, public and private actors assume co-ownership of water assets and co-responsibility for the delivery of water services. The public and private sector partners can either form a new company or share ownership of an existing company (*e.g.*, when the government sells shares in an existing company to the private sector). Joint ventures create a new entity to implement the various types of project structures – for example, the government may award the jointly owned firm a service, BOT, or concession contract.

Joint ventures provide a vehicle for "true" public-private partnerships in which governments, businesses, non-governmental organisations and others can pool their resources and generate shared "returns" by solving local infrastructure issues. Under joint ventures, the government remains the ultimate regulator, but it also is an active shareholder in the operating company. From this position, it may share in the operating company's profits and help ensure the wider political acceptability of its efforts. The private sector partner often has the primary responsibility for performing daily management operations.

Financing Structures

Under the joint venture model, both the public and the private sector partners are responsible for investments. Joint ventures require that both parties accept the idea of shared risk and shared reward.

Box 4. Strengths and weaknesses of joint ventures

Potential Strengths
- Joint ventures combine the advantages of the private sector with the social responsibility, environmental awareness, local knowledge, and job generation concerns of the public sector.
- Under a joint venture, both the public and private sector partners have invested in the company and therefore both have a strong interest in seeing the venture work.
- Full responsibility for investments and operations gives the public and private sector partners a large incentive to make efficient investment decisions and to develop innovative technological solutions.
- Early participation by the public and private sector partners allows for greater innovation and flexibility in project planning and helps ensure that both partners are able to optimise their goals.
- Early dialogue between the public and private sector partners can help reduce the transaction costs associated with more traditional tendering processes.

Potential Weaknesses
- The government's continuing regulatory responsibilities may lead to a conflict of interest in maintaining both public accountability and an eye on maximising returns to the venture.
- Private sector organisations tend to focus on the "bottom line" – governments on the process. These differences can create barriers during project development.
- The early dialogue between the public and private parties involved in some joint ventures may lead to alternative public tender procedures such as direct negotiation, which may create a transparency problem and limit competition.

2.4. *Private operation, investment, and ownership, public oversight – full divestiture*

Complete divestiture, like a concession, gives the private sector full responsibility for operations, maintenance, and investment. But unlike a concession, a divestiture transfers ownership in the assets to the private sector. Hence the nature of the public-private partnership differs slightly. In a concession the governments role is to ensure that the utility assets – which government continues to own – are used well and returned in good condition at the end of the concession and, through regulation, to protect users from monopolistic pricing and poor service. A divestiture leaves the government only the task of regulation, since, at least in theory, the private company should preserve its asset base out of self-interest. In reality though, it has proven necessary to install also some regulatory oversight over the assets. So far, complete divestiture has only been experienced in England and Wales, where the regulator retains important rights to appoint another operator in case a water company fails and more generally limits the length of the licenses under which companies operate.

Financing Structures

As for concession contracts, over the lifetime of the contract, the private sector manager is responsible for all capital and operating costs – including infrastructure, energy, raw materials, and repairs. In return, the private operator collects the tariff directly from the system users. The tariff level is established by an independent regulatory agency, so as to minimise the risk of political interference. In order to prevent under investment, the regulator may also need to scrutinise the utility's plans for renovating or enhancing its assets. In England and Wales the regulator requires utilities to report the serviceability of their assets. Information is the crucial ingredient on which the quality of regulation will ultimately depend.

> **Box 5. Strengths and weaknesses of full divestiture**
>
> Potential Strengths
> - In some circumstances divestiture may be more appropriate than a concession, for example, divestiture by sale of shares or management buyout may produce efficiency gains without involving foreign water conglomerates.
> - Divestiture could help develop local private firms capable of working in water and sanitation.
>
> Potential Weaknesses
> - Divestiture may be politically less viable than other forms of private sector participation, due to ideological and constitutional reasons.
> - The regulatory oversight of privatised water services in England and Wales has proven to be extremely demanding in terms of regulatory capacities and institutional complexity.

2.5. Unregulated private provision – small businesses, community organisations

At the same time that governments are exploring these more formal ways to involve the private sector, areas not currently served by government water systems are being served by private providers. Some are small businesses. Others are community-based water and sanitation associations. Many fall outside formal government structures. Most operate in the poorer, peri-urban neighbourhoods that face major issues of land tenure. All possess a largely untapped, potentially extremely valuable, base of knowledge and credibility for building local water businesses eventually to participate in more formal government procurement efforts.

When water services from governments or private vendors are inadequate, community-based provision can fill the gap. Community-based providers include individuals, families, or local community

> **Box 6. Strengths and weaknesses of unregulated private provision**
>
> Potential Strengths
> - Small business and community-based provision taps into local knowledge, which often results in the more efficient provision of services and protects against misguided investments.
> - Community-based arrangements typically reduce initial investment costs by integrating local resources into the project.
> - Small businesses and community-based provision can provide local residents with a stable form of income, which can improve local economic conditions.
> - Small businesses and community groups are dynamic and often able to respond better to customer demand, resulting in more sustainable infrastructure services.
>
> Potential Weaknesses
> - Two major concerns with small business and community-based provision are coverage and scale. Although often successful in specific neighbourhoods, they can be difficult to expand to a larger scale or be replicated in other neighbourhoods.
> - Governments are sometimes reluctant to support community-based providers because their informal methods of service provision are viewed as illegal and unstable. Similarly, "informal" water firms may be hesitant to increase their visibility and contact with formal government systems.
> - To build up sustainable community-based infrastructure projects takes time. Institutionalising and maintaining those structures can be difficult.

associations. Community-based organisations ("CBOs") can play a key role in organising local collective action. They can then work with international non-governmental organisations ("NGOs") and others to organise and fund the water services. For example, community-based providers may buy water in bulk from the local utility and sell it in their community in buckets. Group taps may also be used to provide service to three to six households using only one tap. Other water options include "communal water point service" where 20 to 30 households install metered taps off the main system and regulate their own water use, paying the bill collectively.

Financing Structures

Small business and community-based provision typically involves low initial costs, as little capital equipment is needed and human capital is available through the local providers. Local knowledge generally allows for the development of least cost solutions, keeping expenses low. For community-based provision, initial organisational and material costs are often covered by NGOs, private charities, official development assistance, the local government, as well as by the community itself. Maintenance costs are covered through user fees.

2.6. *Summary*

Various forms of private involvement assign different roles to public and private parties. Matrix 1 summarises the allocation of these roles across a range of structures. What stands out is that the government (indicated by the darkest squares) always retains responsibility for setting and enforcing performance standards – regardless of the form of private involvement chosen.

Matrix 1. **Allocation of public/private responsibilities across different forms of private involvement in water services**

Legend: ■ Public responsibility　▨ Shared public/private responsibility　□ Private responsibility

	Setting performance standards	Asset ownership	Capital investment	Design and build	Operation	User fee collection	Oversight of performance and fees
Fully public provision	■	■	■	■	■	■	■
Passive private investment	■	■	▨	■	■	■	■
Design and construct contracts	■	■	■	□	■	■	■
Service contracts	■	■	■	■	□	■	■
Joint ventures	■	▨	▨	▨	▨	▨	■
Build, operate, transfer	■	■	□	□	□	□	■
Concession contracts	■	■	□	□	□	□	■
Passive public investment	■	□	▨	□	□	□	■
Fully private provision	■	□	□	□	□	□	■

Source: Yale-UNDP Partnerships Program 1998.

For governments, NGOs and others interested in attracting more private involvement in their water systems, the critical questions are: *i)* what lessons have been learned about private involvement in the water sector; and *ii)* based on those lessons, what steps can be taken to attract more private involvement in water services. These topics are addressed in Sections 3 and 4 of this report.

3. Lessons Learned About Private Investment in Urban Water and Waste Water Services

As described in Section 2, governments around the world are increasingly seeking to involve private firms in the provision of urban water and waste water services. This choice, however, raises concerns in many quarters. What are the tradeoffs governments must consider when deciding whether to provide or oversee water services? Why do some international private investors have trouble finding attractive water deals when there is a strong demand for more investment in water services?

The purpose of this Section 3 is to summarise some of the lessons learned about the factors influencing private investment in water services. Its focus is on understanding the major issues facing both governments and potential private investors. Recommendations for actions that governments and others can take to meet these needs are provided in Section 4.

3.1. *Private involvement does not relieve the government of its responsibility to ensure that basic rights to water are met*

3.1.1. Governments are responsible for ensuring that their public's basic needs are met – nvolving the private sector is just one tool for doing so

To the extent governments have taken on the responsibility for providing water to their citizens, they have a choice in how to do so. They can provide those services directly, retaining complete control over all aspects of the operations. Alternatively, they can involve private parties, to a greater or lesser degree, if it is likely to increase efficiency, access to technical and managerial expertise, or private investment. Either alternative may succeed or fail depending on a wide range of factors.

Either way, governments retain the ultimate responsibility for ensuring that basic public needs are met. Private sector involvement does not mean that governments are relieved of, or must abdicate, that responsibility. Rather, it means that governments must change the manner in which that responsibility is met.

3.1.2. If governments decide to involve private firms to help meet their responsibility, they also need to shift from being the manager of the water system to its overseer and regulator

As the provider of water services, the government manages all aspects of the water system. It decides what is to be built, who is to be hired to do what, how much is to be charged, what quality of water is to be provided, and all related matters.

The government takes on a very different role if it decides to involve the private sector. At least for those tasks assigned to the private firm (and they may be small or large), the government stops being the day-to-day manager of that work and becomes the overseer of the work done by the private firm.

Making this shift from provider to overseer and regulator is extremely difficult for many governments. As noted below, however, the government's regulatory capacity is one of the most critical considerations for potential private investors. If the government's regulatory and policy capacity is weak, little international private capital will flow into the water sector. The only options will be short-term, low-cost management contracts, or domestic investors.

3.1.3. Decentralisation further complicates the shift in government roles

At the same time that governments are considering increased private involvement in water services, many are also engaged in broad-ranging decentralisation of functions from national to municipal authorities. This can range from responsibilities to supply services to the authority to collect new types of revenues.

Decentralisation often has both positive and negative implications for efforts to increase private involvement. Positive aspects can include bringing responsibility for ensuring service provision closer to the users. This should increase the influence of customers on the quality of the services rendered and the prices charged. The biggest potential problems associated with decentralisation concern the ability to attract new private investment. First, the process of decentralisation is often associated to new uncertainties. Second, the local level often lacks the capacity to deal with complex forms of private involvement. Third, decentralisation might result in areas too small to support the often significant transaction costs associated to private sector involvement.

3.2. Water presents a paradox for many private investors – great opportunities, great risks

At the same time that the water sector poses complicated choices for governments, it also presents two faces to private investors – one very attractive, the other raising major concerns.

Three major characteristics of the water sector make it most attractive to private investors: the need for expanded water services; the revenue streams created by the (relatively inelastic) demand for water; and the opportunities for increasing those revenue streams through improved system performance.

3.2.1. *There is a great need to expand water services*

The global scale of the need for new investment in water and wastewater services is staggering. World-wide, only 200 million people are receiving water services from private sector operations so far. Combined with the growing international private investment in developing and transitional countries, and the increasing decentralisation of responsibility for urban public services, large opportunities are clearly presented. These opportunities are of great interest to many potential private investors, most active are the French and English private water companies.

3.2.2. *Users are willing and able to pay for many water services*

Since access to drinking water is a basic need, it has great value to individuals. Most urban dwellers already pay something for their drinking water – either through connections to formal, networked systems or purchases from informal vendors and community-based providers.

As a result, the potential revenue streams are sufficient to interest private investors in drinking water services over the long-term. Even poorer, non-networked urban neighbourhoods can be viewed as reliable sources of revenue, as they often pay more for their drinking water than wealthier areas.

More difficult issues arise, as discussed below, for other parts of the water cycle – particularly wastewater collection and treatment. While people are often willing to pay something to have sanitary wastes removed from their residences, they often value it less than access to clean drinking water. Even less consumer value is placed on treating sanitary wastewater once it is taken away.

3.2.3. *Private involvement often improves system performance – and increases revenues*

As discussed in Section 2, governments often seek to involve the private sector to improve system performance – which it often does. Some of these improvements are due to the commercial incentives facing private providers, leading to a focus on greater efficiency and customer satisfaction. Other improvements stem from their greater access to technical and managerial knowledge. Finally, private providers may be able to tap additional sources of private capital – from their own equity to commercial debt – to support system operation.

3.2.4. *Water is a basic human need – and an economic good – a volatile mix*

People have a right to water. However, this right does not entitle one to an unlimited amount of water – due to ecological, economic, and social constraints. Governments exist to help citizens meet their basic needs. Water is such a need, so governments have traditionally sought to provide it. Many politically powerful citizens have come to view low-cost water as an entitlement. Many government officials have increased their power by controlling water services and maintaining low prices.

At the same time, water is increasingly seen as an economic good. It has financial value to customers, and there are significant costs involved in supply and treatment. Given the scarcity of water in many areas, there have to be incentives to conserve it. Water should be allocated by prices, in combination with regulatory action. It can generate revenues sufficient to support private investments.

This is a volatile mixture, inviting major disputes over competing values and pricing. The bottom line, however, is that governments need to ensure that basic human needs are met. They have to remain involved in the water sector, even with private investment. The question for potential private investors is whether the form of the government's continued participation makes the investment more or less attractive compared to other opportunities. Much of the answer will depend on the clarity and predictability of the government's oversight.

3.2.5. *Water networks are long-term, inherently risky investments*

Clarity and predictability of government involvement are particularly important to private investors in networked systems – piped drinking water distribution or waste water collection.

Networked water systems have extremely high capital costs, well in excess of those in many other infrastructure services. They are mostly financed with debt, for as long a term as is commercially available. Given the high initial costs, extremely long payback periods are necessary. Revenue streams need to be as secure as possible, free from disruption either from governments or from water users.

The following characteristics increase the risk to potential private investors in networked systems: the scale of the capital costs; the amount of debt necessary; the long payback periods; the government involvement in prices, standards, and collateral; as well as the potential for currency fluctuations.

3.2.6. *Water fees are often too low to support major private investments*

Many governments currently sell drinking water for prices well below the cost of providing the service. In some cases, this is to ensure that the basic needs of all citizens are met, even those who find it hard to pay. In other cases, it is to build political popularity or to avoid the civil unrest that might accompany efforts to increase prices. Even lower prices are usually assigned to sanitation services and raw water abstraction. In each, it is the political, not the economic, value of the water that is driving the calculation.

The impact on potential private investors is clear – the lower the revenue stream, the smaller the investment they will be willing to make.

3.2.7. *Costs and risks are often too high*

High capital costs, low fees – these are often just the beginning of the list of risks facing potential private investors in water. Other major areas of concern include high, up-front transaction costs, project-specific risks, and country-specific risks.

- Transaction costs

The more money private investors have to put into a deal, the more they will have to take out in order to make an acceptable profit. Investors start incurring costs when they start looking for deals by speaking with municipalities. The costs continue to mount through the due diligence, bidding, and contract negotiation phases. All of these costs are incurred before any investments are made in improving system performance. All of them have to be recovered from project revenues – either from government payments or water user fees.

In addition to increasing project costs, one effect of high transaction costs is to focus potential international investors' attention on the biggest cities. It is there that they are likely to find revenue streams large enough to justify the up-front costs. This only compounds the problems facing smaller cities seeking to bring private sector expertise to bear on improving their water services.

- Project risks

During pre-investment due diligence, potential international investors in water systems seek to identify, quantify and mitigate two major types of risks: project and country risks. As their names imply, project risks are those facing an individual project, while country risks are those facing any project in an individual country.

In the water sector, many project risks stem from the choices made by government. Will performance standards be changed over the life of the contract? Will capital investment requirements be changed? How will prices be adjusted? Are the procedures for making any such changes likely to be applied in a fair and predictable manner? Will the government make the payments to which it has committed? Depending on the length of the contract, the scale of the contemplated private investment, and the source of the revenue stream, project risks can pose major barriers to private involvement.

- Country risks

The country risks facing water projects can also be large, depending on the scale of the investment sought. Currency risks – particularly in projects carrying large foreign currency debt – are a major concern. Macroeconomic conditions and the local political climate can pose major issues. The clarity and predictability of legal and administrative frameworks, particularly those concerning foreign investors, are important factors for international companies. Governments are in the best position to reduce such country risks. Whether they are able or willing to do so varies dramatically from country to country.

3.2.8. Governments and users are often not willing or ready to address risks to investors' satisfaction

Clearly, private investors must take responsibility for many risks. This is particularly true for the business risks that they are in the best position to manage, such as construction costs, treatment plant performance, or the efficiency of billing and collection activities.

As noted above, however, other significant project and country risks lie more within the control of governments and users. How they respond has major implications for the willingness of and the terms on which private firms will choose to make investments.

- Issues for governments

Many governments find it hard to address these risks to the satisfaction of private investors, in part, because they raise fundamental issues about the roles and capacities of the public sector.

At the heart of the issue for many governments is crafting their new role in the water sector. As discussed above, the fact that water is a basic need justifies continued government involvement. At the same time, in order to attract private investment and capture the opportunities for improved performance, governments need to transfer some of their traditional functions to the private sector.

In addition to these fundamental questions, other politically charged issues often affect government action in the water sector. The two most common are concerns about price increases and labour issues. If governments have traditionally under-priced water services, and now seek to increase prices in order to support additional investments (public or private), considerable public opposition often needs to be overcome. Similarly, if private operators seek to increase efficiency through reductions in the number of employees, substantial opposition from labour interests can be expected.

Finally, private involvement requires many governments to acquire other new skills. For example, as municipalities take on new responsibilities under decentralisation programs, they find themselves negotiating multi-million dollar contracts with private companies. For many, this is a new experience. Often, the results are major disparities in bargaining power, particularly when large, international water operating companies are involved. These problems are only magnified the further one goes from large, capital cities.

Issues for users

For many existing customers of government water systems, the potential involvement of private providers is viewed as a mixed blessing. On one hand, their service is likely to improve. On the other, they are likely to pay more. This leads to many questions: Why cannot the government do better in providing this social good? Is it appropriate to make a profit from meeting a basic human need? Unless these and related questions are answered to the satisfaction of these, sometimes politically powerful, existing customers, there will be a lukewarm or even hostile public response to efforts to involve public investors.

3.2.9. *International private water operating companies are limited in number and cannot do everything*

Most of the private investment in water services to date (particularly in developing countries and transition economies) has been made by organisations at either end of the spectrum of private providers: 1) a small number of very large international private water companies; or 2) a huge number of very small, informal water vendors or user organisations.

Given most governments' goals of attracting increased technical and managerial experience, as well as potentially large sums of new private capital, their almost exclusive focus has been on involving the large, international companies. These firms are viewed as "one-stop-shops" for meeting all of the governments' needs. Efforts are made to lock them into long-term investment and operating contracts.

While this approach can work exceedingly well, it also restricts the potential scope of private investment. The international water companies are few in number. They do not have an unlimited capacity to make investments. They will seek out and concentrate on the largest, most potentially profitable opportunities. By definition, this means that investments in poorer or smaller service areas are often left out or delayed. New ways need to be found to involve a greater number of private investors – of various sizes, nationalities, and experience – in improving the performance of water services.

Recommendations for steps that governments and NGOs can take to learn from these lessons and increase private involvement in urban water and waste water services are provided in the next Section 4.

4. Steps Governments and Other Actors Can Take to Increase Private Investment in Urban Water and Waste Water Services

"It is not the money that is missing, it is the deals" – a lament heard from many would-be international private investors in water and wastewater services. Whether true or not – the fact that many potential investors feel this way creates tremendous opportunities for interested governments, non-governmental organisations and others to offer attractive investment opportunities to willing investors.

Developing attractive investment opportunities requires work across three major areas: market frameworks; information; and shared investments. These efforts should focus on applying the lessons learned about private investment in water described in the above Section 3. Recommended steps for doing so are set forth below.

4.1. *Adopt market frameworks that encourage private investment in water supply and sanitation while protecting the public interest*

Governments make the markets in water. They usually hold the basic ownership right to water. Through regulatory structures, they allocate rights to use water, protect water quality, and control the price and performance of networked water systems.

Governments should use their control over the water market to provide sound foundations for both private investment and protection of the public interest. In doing so, they should adopt and implement market frameworks that accomplish the following tasks.

4.1.1. Get prices right

Water should not be provided for free. Water fees should reflect the costs of providing the service, including profits when private investors are involved. Legitimate government concerns over ensuring that all people have access to water and waste water services should be addressed separately from the efficient operation of water services. In fact, for many poorer areas, connections to the networked system will reduce, not increase, the amounts currently paid for informal water provision. Lifeline or minimum service rates can be set, sometimes cross-subsidised by higher volume users, although such a system should be clearly delineated and recognised in accounting statements. Separate mechanisms for income support or assistance in paying water bills can also be established (see below). The prices that need to be "gotten right" are not just for drinking water – attention should also be given to prices for wastewater treatment and raw water abstraction at the same time (see discussion below).

4.1.2. Set performance standards to reflect local needs and demand

Market failures require that governments set and enforce performance standards for many parts of the water cycle. Customers of monopoly suppliers of drinking water rely on governments to control drinking water quality, quantity and price. Similarly, environmental advocates and raw water users look to governments to set and enforce standards for pollutant discharges to surface and ground waters. The levels at which any of these standards are set have major cost implications. Therefore, performance standards need to strike a difficult balance between the need to protect customers and the environment, and the need to maintain water services at an affordable level.

4.1.3. Improve regulatory capacity

Even when private investors are involved, governments retain large amounts of control over water and waste water services through their responsibility to set and enforce standards of performance (economic, environmental and social). As described above, this is primarily aimed at protecting customers against abuse of monopoly powers and water sources against further pollution. Any government regulator has to strike a difficult balance – being sufficiently engaged with the regulated firm to be confident in receiving the information necessary to fulfil regulatory responsibilities, but without unnecessarily interfering with management decision-making. Finding the balance calls for clear rules, fairly applied in a predictable, timely and transparent manner. Involving users in regulatory decision-making can help improve regulatory performance.

4.1.4. Choose methods for private involvement that fit local needs and context

Governments do not choose to invite private investors into water services just so that the firms can make profits. Rather, the purpose is to improve the delivery of water and waste water services in ways the governments believe they cannot achieve by acting alone.

Many different forms of private involvement are being tried (see Section 2). There is no universal "right answer" on how to use private investment to help improve water services. Ultimately, however, governments, investors, and users need to devise arrangements that address their local needs in a manner that fits the local context. In the particular NIS context, where the private sector is hesitant to engage, it might be appropriate to start with those methods for private sector participation that involve a low risk for the private operator (*e.g.*, a service contract), and later on use the experience gained to move towards more ambitious forms of involving the private sector.

4.1.5. Use controls over market access to encourage creativity, competition, and inclusion

Since governments make the markets for water, they also control access to them. They should use that control to expand even further the potential benefits from private sector involvement, by stimulating competition between different private providers, and between public and private providers, while ensuring a level playing field.

- Use bidding procedures that encourage innovation in system design

Traditional procedures for public procurement assume that the government knows the technical specifications of what it wants. A very different situation exists when trying to optimise water and waste water services across large metropolitan areas. Many technical options exist, along with many different approaches to managing system assets. Working to traditional government design specifications may well not result in the most cost-effective solutions.

In these circumstances, the bidding process should be used to generate innovative designs and management approaches reflecting the commercial experience of the private bidders. For example, bids could be based on the costs of meeting specified performance standards within prescribed periods – leaving the bidders free to determine the most effective methods for meeting the standards. Such bidding procedures will require governments to modify their approach to technical design issues.

- Expand access to the formal market to more potential bidders

There are limits to the number and types of projects the handful of big, international water companies can take on. Other types of companies – alone or in combination – can also provide the necessary expertise. For example, operation of treatment plants can be separated from billing and collection, which can be separated from management of project finance. Small, local companies are more likely to be capable of bidding on small projects or in smaller cities, and their potential involvement has the added benefit of achieving local business development goals. All of these potential private investors need to be brought into the effort to meet the huge demand for improved water services. All can help increase the competitive pressures for improved performance.

4.1.6. Deal with labour issues early on

The employees of the existing public water utilities may be apprehensive of the involvement of the private sector, even though experience has shown that the operators tend to retain a large proportion of the existing workers and wages usually go up.

Even so, the very purpose of private participation is to enhance productivity, which often will mean doing a better job with fewer people. Many tasks currently performed by public water utilities should be outsourced. Even without the involvement of a private utility operator, the Government will sooner or later be confronted with tough labour decisions and will have to balance the interests of the workers with those of the entire population demanding better services. As part of the process to involve the private sector in the provision of water and wastewater services, social safety plans should be prepared.

4.1.7. Include water in efforts to improve domestic markets for private investment

In parallel with any efforts to attract private investment in the water sector, most governments in developing or transitional countries are trying both to strengthen their local capital markets and to increase their country's attractiveness to foreign investors for all types of projects. Improving the performance of local banks, building local regulatory frameworks for bonds and other long-term debt financing, and expanding local stock exchanges are all important parts of the efforts to increase access to local currency financing. Since most water revenues are in local currency, increasing access to longer-term local currency financing will help reduce currency risks. Reducing and improving the transparency of regulatory controls on foreign investment is also important.

4.2. Provide better information

In addition to building effective market frameworks, governments and other actors can help improve the efficiency of water markets by making better information more readily available to all stakeholders. This is particularly important for increasing the capacity of users and governments to take on their new roles, reducing transaction costs, and increasing competitiveness pressures.

4.2.1. Increase public awareness of and capacity to influence water issues and trade-offs

Users are only willing to pay or pay more if they understand the needs and the benefits. Providing them with information on competing systems and product quality, local water issues, the options for addressing them, and the costs of doing so is critical. Such information will help prepare users to: 1) participate in the decision-making process on the standards to be met; 2) contribute to the costs of meeting those standards through payment of water fees; and 3) help oversee the selection and/or performance of the private firms.

4.2.2. Expand municipal capacity to work with private water companies

Many municipal officials are being called on to undertake complicated, new roles – negotiating contracts with international water companies, regulating the private delivery of water services, participating in the financing for water projects. In many cases, they welcome advice on how to best fulfil these roles in manners consistent with their public responsibilities.

Peers in other municipalities are one of the best sources of such advice (several programs are now underway to help meet this need, such as the UNDP's Programme on Public-Private Partnerships for the Urban Environment). Local pools of professional advisors – financiers, lawyers, engineers – familiar with the issues facing private investment in water services also need to be developed. In the transition period donors and IFIs can help to make advice from international experts available to municipalities who otherwise would be unable to afford these services. Governments should ensure that they are advised by internationally recognised experts when they negotiate with private companies.

4.2.3. Reduce transaction costs by sharing or requiring less information

Some of the large transaction costs related to private investment in the water sector are unavoidable. It should be possible, however, to reduce transaction costs by sharing some types of information. At a global level, this might include publicly available databases of water projects and individuals involved, standard form bidding and contract documents, standard statements of qualifications, and similar items. For specific projects, it might include shared data on the state of the current water system or on customer willingness and ability to pay, to avoid duplicative efforts to collect this information. In fact, customer surveys may have greater credibility if done by third parties such as NGOs or academic institutions.

4.2.4. Enhance competitiveness pressures by reporting on performance

While it is difficult to introduce ongoing competition into networked water systems, it is possible to simulate aspects of competition by comparing performance across water providers working in different localities. Often referred to as "yardstick competition", such information programs are built around common performance indicators. Periodically, the performance of different water providers is assessed against those indicators and the results released to the public. No company or government body wants to be at the bottom of such a list. Not only can it add to user complaints, it may affect the firm's ability to raise additional finance. Considerable pressures to ensure a good ranking through improved performance are created.

4.3. Make shared investments in urban water and waste water services

Because water is a basic need and an economic good, both governments and private firms have to invest time and resources in its provision. At a minimum, governments need to invest in regulatory systems sufficient to ensure that public goals are met and private firms need to invest to provide quality service. Other, shared investments may also be appropriate, aimed at reducing upfront and transition costs, as well as to help ensure that social goals are met. Applying public funds alongside private capital can be an effective technique for achieving these goals. While the debate over the use of public funds to support private activities is beyond the scope of this paper, some areas for consideration are suggested below.

4.3.1. Share the costs of developing pre-investment information

Much of the information affecting transaction costs (discussed in section 4.2.3 above) is important to both sides. Private investors will have to develop some of it for themselves in order to be confident in their investment decision-making. Other types of information – such as global databases, form documents, surveys of existing system performance and customer willingness to pay – can be developed for joint use, applying funds from both public and private sources.

Even more fundamentally, water districts or municipalities in developing and transition countries often do not have the resources to prepare for bidding procedures, or to evaluate unsolicited proposals. Donor organisations can play a role in helping to promote open bidding and informed evaluation of private investment, creating "study funds" that water districts and municipalities can draw from for feasibility studies, project documents, as well as packaging and evaluating privatisation proposals that clearly reflect their needs, standards, and willingness and ability to pay.

4.3.2. Provide separate income support

Ensuring that all citizens have access to clean water – regardless of their ability to pay – is a key goal for most governments. To the extent that governments are looking to subsidise the costs of water for the poor in some way, they should not do so through universal lower fees for water use. Rather, they should separate income support from water rate levels using techniques such as "water stamps".

4.3.3. Ease transitions

If water rates need to rise or labour costs need to drop as part of a move to improve water services, public funds can help ease the transitions. Governments can choose to provide a declining universal rate subsidy over time to ease the transition to an economic price level for water. Similarly, governments can provide retirement, relocation, retraining, and similar packages to public employees affected by the shift to private investment. These types of public support – limited duration, declining amounts – can promote, rather than impede, the move toward greater efficiency in the water sector.

4.3.4. Manage political risks

Given the level of government interest and involvement in the water sector, one of the greatest areas of risk for private investors is how governments choose to manifest their participation. Will they keep their promises? Will they change the targets?

The best protection is for government officials to understand and honour their side of any bargain. There must be security both in terms of political structures, contract dispute mechanisms and political attitudes to tariffs. There must be confidence that *a*) water laws and regulations are in place that support economic efficiency goals, and *b*) the tariffs needed to finance investment, operations and profit are achievable.

International donor agencies can help reduce political risks. By investing in a particular project, they provide an attractive "halo effect" for potential private investors. Thereby IFIs would help to develop demonstration projects that will make obvious the advantages of private sector participation to governments and potential investors.

4.5. Conclusion

Finding the right combination of roles for public and private actors in urban water and waste water services starts from a few basic principles:

- Governments control water markets.
- Users pay.
- Businesses need to make a profit.
- Donors can help.

How these principles are best applied to any particular situation depends on local needs in the local context. They do suggest the following, however:

- Governments should not view private involvement as reducing the importance of their role. In fact, the roles played by governments are among the key determinants of both how attractive the opportunity will be to potential private investors and how effective private involvement will be in serving the public interest over time.

- Users have to be willing to pay for the water and wastewater services they receive, whether or not these services are provided publicly, privately or jointly. They need to understand the water issues facing their city and the options for addressing them, including the costs. User preferences should be reflected in the choices made by governments affecting the water services, particularly when users will be asked to bear the cost. Their preferences should also inform the oversight of private service provision, either directly as part of the regulatory structure or indirectly through solicitation of data on user satisfaction.

- If private investors are involved, governments and users need to accept that they have to make a profit – at least as high or higher than that offered by competing investment opportunities. The returns realised by the municipality in terms of increased efficiency of service delivery should be enough to cover the incremental costs of the profit – if not, the private firm is not doing its job, or its involvement was not properly designed. Private firms should also recognise that their business interests are best met by taking a "reasonable" return over the longer term, rather than large profits in the short term which bear a high political cost.

- Finally, donors can help – or hinder – these efforts depending on the approach taken. If they support government efforts to put effective policy frameworks in place, including the involvement of local users, they can render a real service. Donors are also well positioned to collect and make available experiences, contacts and information on the different ways governments around the world have chosen to involve private firms. To the extent possible, this information should be freely shared across programs – there is more than enough work to be done. Donors are obviously less helpful if their assistance is designed to impose a standard model approach on each, different situation.

All of these parties are part of the recipe for successful – as measured by all of the stakeholders – private involvement in urban water and waste water systems. By understanding, respecting and reflecting their goals and needs, the private sector can play a constructive role in improving the delivery of water and wastewater services.

Chapter 6

WATER SUPPLY AND SANITATION THE PRIVATE SECTOR'S VIEW ON RISKS AND OPPORTUNITIES

General Introduction

The private sector enters new markets and makes investments routinely in order to obtain access to more consumers, to expand market share, and to earn income streams covering the cost of the capital investment, covering operation and maintenance expenses and producing a profit. Historically most business involvement and investment has been domestic where business leaders are very familiar with the framework conditions including laws and regulatory requirements.

When businesses enter or invest in foreign markets, they need to assess the conditions in the host country. Obviously, business will be hesitant to invest in any foreign nation which:

- is in the midst of a civil war or state of anarchy;
- has an inadequate legal, civil and commercial code;
- has a track record of expropriation of private investments.

Some businesses will even enter markets and invest under these extreme circumstances. However, they will only do so if there is a very high risk premium – for example, the investment can be recouped in one to three years; the rate of return on investment is very high; International Financial Institutions or other governments provide guarantees.

It is instructive to note that well over 90% of all Foreign Direct Investment (FDI) takes place between OECD nations or with 13 other rapidly developing nations. Thus the vast majority of nations attract a very small portion of global FDI. The message here is that many countries have been unable to attract foreign direct investment for a wide range of reasons. The markets may be too small; the average per capita income may be too low; there may be social or cultural restraints inhibiting private capital investment.

So far, the NIS have been among those geographical regions which have only attracted only limited foreign investment in the water sector. Out of total FDI flows of US$25 billion between 1990 and 1997, only 6% went to the CEEC/NIS region, and probably less then 1% to the NIS alone (Figure 1).

However, there is an abundance of capital in developed markets, primarily in the form of personal savings and pension funds available for investment. Business will go wherever there is an opportunity for profits adjusted for a wide range of risks. It is also important to note that different companies will evaluate these risks differently.

The business community regards safe drinking water and basic water sanitation as one of the prerequisites for deciding when to invest in a city or economic region. Thus good water service is a necessary, if not sufficient, condition for sustainable development. NIS government can attract more FDI when fundamental water infrastructure is in place and maintained and the public health of its workers and customers is assured. The private sector can help NIS governments achieve these objectives by offering technological and managerial expertise, as well as investment financing.

Figure 1. **Regional share of FDI in the water sector**

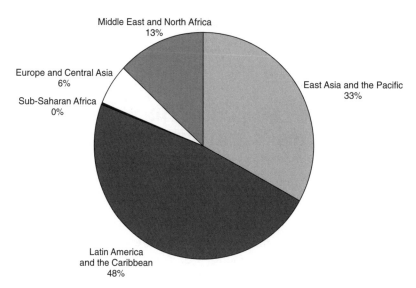

Source: World Bank, Public policy for the Private Sector, Note No. 147, 1998.

Investment in Water Supply and Sanitation

The first question to ask is why does any nation want or need the private sector in their domestic water service business? The best answers were presented by the World Commission on Water and reaffirmed by the Ministerial Declaration at the World Water Forum in The Hague in March of 2000:

1. *"... it is clear that the private sector can considerably improve the dismal technical and financial performance that characterises most public utilities in developing countries."*[1]

2. *"Governments know they don't have enough money to double investments in water services to $180 billion a year." – Minister Eveline Herfkens – The Netherlands.*

There appears to be a growing consensus that the way forward to improve water services in both NIS and developing countries will involve a wide range of new **Public – Private – Partnerships.**

However, there are significant differences between NIS and most developing country markets. In the NIS there has been a long history of significant investment in water supply and sanitation infrastructure by central government authorities. For a wide range of reasons, much of this infrastructure has not been well maintained. Pumps are not functioning and need to be replaced. Pipes and valves are leaking; water wastage rates are abnormally high. Historically, citizens regarded water services as a free good or one provided by government at an extremely low charge. Unfortunately these low water charges encourage excess consumption and, whenever governments are short of funds, there is deferred maintenance with no modernisation of basic delivery systems.

It is clear that more resources must be made available to rehabilitate deteriorated water infrastructure in the NIS. The routine management and customer service also must be improved. The private sector contends that since they are specialists in water service, they can out-perform public sector service organisations in providing these services. Private sector water companies provide quality service with significantly fewer employees per thousand customers served. The quality of the water delivered consistently meets strict European Union or equivalent drinking water standards and stringent emission limitations for wastewater discharge. Private sector operators are monitored by independent government regulatory agencies and service contracts are subject to competitive bid at specified time frames.

Water supply and sanitation is essentially a service industry. There are no products (other than pure drinking water), no export markets, no opportunity to earn foreign exchange credits. Thus unlike minerals, metals, forest products or general manufactured goods, the manager of a water service business must recoup costs through user charges or through some form of government contracts. These fees will almost always be collected in local currency. Thus the foreign investor will be looking for some form of premium to cover risks associated with potential currency devaluation *vis-à-vis* the capital investment normally denominated in "hard" currency widely traded and exchanged in international markets. Rules restricting the repatriation of legitimate profits discourage private sector investment by foreign corporations. Investment in water infrastructure is capital intensive; even some maintenance programs require significant capital investment. It is for this reason that water investors seek a stable regulatory system which will enable them to recover these long term investments over the life of the capital stock.

There is no one perfect model for new public-private water partnerships. One can envision models ranging from almost total public ownership and operation – the historic model – to almost complete private ownership and operation. Even in the extreme case of virtual total privatisation – *e.g.* the English model – the role of government regulation, oversight, and rate setting remains crucial.[2] This is just as important to the private sector as it is to the government, because private sector seeks a predictable working environment. Government's regulatory function should have two main objectives:

1. to protect the public interest – to prevent monopoly extraction of excessive charges for services so essential to life and well being; to ensure the quality of the water in order to protect public health and the environment; to monitor performance and enforce contracts;

2. to protect the private investor by ensuring his revenues cover his legitimate costs, enforcing payment for service and preventing theft of the resource.

Alternative models could include various options to build, own and operate specific portions of the infrastructure; be limited to a management or other franchise. No matter which model of Public – Private - Partnership is negotiated, there is a long list of risks which businesses consider before actually entering new markets and investing.

Potential Risks from the Private Sector's Viewpoint

- Political interference in normal business decisions – this should in no way impede the negotiation of the core contract; what should be avoided is constant renegotiation and second-guessing of the basic decision agreed upon after fair negotiations.

- Day to day "micro-managing" by public regulatory bodies – private operators must be given a degree of autonomy to act within reasonable guidelines without bureaucratic review. In no way should this limit legitimate oversight and measuring against agreed upon standards.

- Regulatory uncertainty – constant stops and starts or changes in policy direction.

- Socio-economic instability which might impede due process – radical swings in macro-economic policy which result in significant currency devaluation, dislocations, widespread unemployment or loss of purchasing power of citizens – customers.

- Labour relations which prevent operation of sound business practice – unauthorised work stoppages; work to rule agreements which preclude continuous improvement in worker productivity and efficiency in operations.

- Lack of willingness to pay by some customers and lack of political support for acceptable fee collection or enforcement procedures.

- Lack of transparency in the process which creates suspicion among key stakeholders that their interests are not being protected.

- Too high expectations by some stakeholders – this is the "something for nothing syndrome"; individuals usually will pay some reasonable fee for safe delivered water; often they are less willing to pay for sanitation services because they do not see the immediate benefits – those who live up-stream are unwilling to pay to protect the interests of those living down-stream; further pollution of underground aquifers may be invisible and unnoticed until serious health effects result; finally there may be some reluctance to pay for environmental protection when the benefits are diffuse and unclear.

- Non-payment by government or major industrial customers – large users may expect discounts for quantities used or the economic system has never charged them for this input cost of production; there may be a tendency for commercial interests to use political influence to avoid paying their fair share.

- Failure of legal system to resolve disputes on a timely basis.

In most NIS many of these risks are still perceived to be important. In the EBRD investment climate survey, which includes many of the risks listed above, the NIS scores are well below those of leading Central and Eastern European countries (Figure 2). Unless these risks are reduced and the investment climate improved, business will continue to show little interest to invest in the NIS.

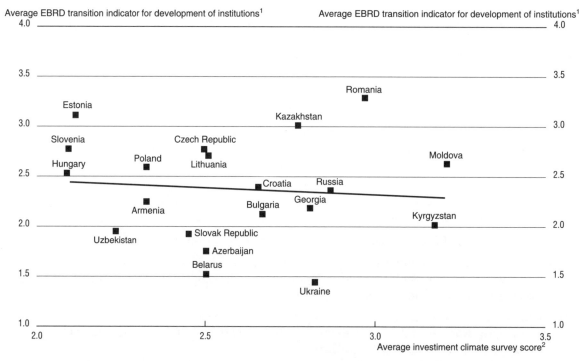

Figure 2. **EBRD transition indicators and investment climate survey scores**

1. EBRD transition indicators for the development of institutions comprise the scores for banking reforms, non-bank financial institutions, competition policy and enterprises. The EBRD transition indicators range from 1 (little progress) to 4 (substantial progress).
2. Investment climate scores are average ratings across the dimensions of macroeconomic stability, taxation and regulation, law and order and the judiciary, as well as finance and infrastructure. Scores can reach a maximum of 4. A higher score means a less favourable investment climate.

Source: EBRD, Transition Report 1999.

Benefits to the Public Sector of Public – Private – Partnerships

- Access to private capital for new infrastructure and for maintenance of deteriorated systems.
- Improved operational efficiency based on best practice and innovation from a water service provider with a proven track record – these companies are dedicated to water issues and, unlike municipal governments, are not trying to address a myriad of social services – water service is their core business and they succeed only by being better than other competitors both from the public and private sector.
- Delegation of this task allows local governments to focus attention on other local services – health and education.
- Properly designed and managed water supply and sanitation systems are a necessary if not sufficient condition for sustainable development – essential infrastructure which is needed to attract more investment by other business sectors.
- Proven ability to deliver capital construction programs on time and within budget.
- Ability to maintain current system, expand system and improve fee collection systems.
- Proven tack record in complying with safe drinking water standards and in addressing water quality and other environmental issues.

Benefits to the Private Sector of Public – Private – Partnerships

- Secure moderate but long term profit stream.
- New customer base allow for continued economic growth of the enterprise.
- Ability to sell-on other related products and services.
- Increased economies of scale – ability to evaluate and buy the best equipment from vendors.
- May lead to "multi-utility" business opportunities.

Finally there is one other benefit accruing not to private sector operators solely, but to business collectively. Improved water supply and sanitation infrastructure enhances the probability that this market becomes a region where sustainable development can thrive. The municipality with a good water service system is a better place to do business, to build new factories and to market new products. A primary reason why business, in general, supports improved water infrastructure is that such service contributes to poverty alleviation and fosters new markets for sustained economic growth.

Key Conditions for Successful Contracts

- Transparent and well defined bid/award process with achievable and affordable performance targets – industry prefers clear and reasonable targets so that its performance can be measured and assessed objectively.
- A sound regulatory mechanism – Municipal and National Government Capacity **Building is essential (this is one reason why many advocate using limited grant money to build up local capacity to negotiate confidently with the private sector) – local authorities may feel more comfortable during the negotiation process if they have access to independent experts or consultants to provide technical and commercial advice.**
- Legal framework in place to settle any disputes expeditiously and build public trust in the system – this is another way of reducing inappropriate political pressure in the system to provide free or "below-cost" service to favoured customers.
- Proactive policy concerning low income customers – internal subsidies within the rate structure is one option; direct subsidies by local governments to cover costs of basic service is another option; the government can assume responsibility for meeting minimum service levels for those unable to pay.

- For NIS the legal status of vodokanals will need to be clarified – who is responsible for existing indebtedness? The property rights between municipalities and the vodokanals including areas of responsibility must be clarified.
- Accurate assessment of existing water infrastructure with honest service data – if the inherited system is near collapse then there needs to be clear decisions on the timetables and costs for upgrading or reconstruction.
- Reliable water demand forecasts and reasonable timetables for expanding system to reach those not currently served – avoid unrealistic expectations – involve the public in a transparent discussion of the levels of service anticipated.
- Clear agreements on the timing and budgetary implications of investments in up-grading, maintaining and expanding infrastructure.
- Building positive relations with existing and future workforce – proper training and clear assignment of responsibility – clear rules for separation of non-performing employees – no "featherbedding" – since many public sector organisations have excessive numbers of employees when compared with private operators, there should be a strategic plan for dealing with redundant workers in a fair and equitable manner.
- A plan for managing a successful transition from the existing service organisation to the new Public – Private – Partnership arrangement.

Involving stakeholders

One way to reduce all of the above risks is to involve all of the key stakeholders in a transparent process. **The most important stakeholders are the customers** – the businesses, the individuals, rich and poor, who will be the beneficiaries of improved service. Their involvement is crucial since ultimately they must pay for the services received. The private sector is anxious to work with all stakeholders since the success of their investment depends on each of these groups. Other stakeholders include:

- Customers – the most important stakeholder as cited above.
- The public partner or asset owner – either the city, region or national government – ultimately some government regulatory authority retains ultimate control of water and water services to ensure that the public interest is protected.
- The private operator – this is the business that invests its management skills, know-how, technology and financial resources into creating a viable and sustainable water service system.
- The banks or other institutions providing investment capital – often the water service operator obtains the capital needed to upgrade, maintain and expand the system from 3rd parties. These "bankers" have their own criteria for where and when to lend. If a project proposal is not "bankable", it simply does not go forward.
- The workforce – the designers, construction engineers, maintenance staff, environmental and health staff and the accountants, auditors and bill collectors – who ultimately determine how well the system functions. Here there is a potential conflict of interest between those who view the system as a source of employment versus those who want service at the lowest possible price. Often this poses difficult political choices for political decision makers. In the longer term, overstaffing by either a public or private water service organisation cheats the consumer by imposing unnecessary extra costs.
- The regulators – these are the trained experts in rate setting and in monitoring performance who act as the agent of the government partner to balance the needs of the public against the needs of the service provider. It is important to partially insulate water service from the passions of the political process, making it more of a technical and business issue but with full oversight by a competent regulatory authority.
- There can be other levels of government who have a stake in the water management of a river basin or region which extends beyond the service area in question.

- Vendors and suppliers – they can be local or foreign suppliers of key components of the water system – they will be interested in the design of the system and in an open and transparent bidding system as suppliers to the operator.

Often the process of involving the stakeholders can be long and tendentious. However, the investment of time at the beginning of the process can pay great dividends as key decisions are made. Essentially what is being described is a confidence building process so that all of the interested parties ultimately agree and accept core decisions.

The emergence of an internal domestic consensus on the level of water service, the technology to be employed, and the allocation of the costs to pay for the service is the best guarantee of longer term success. Customer dissatisfaction is then minimised; political risk to the decision makers is then reduced; economic viability of the private sector operator is then made more probable; and there is a greater chance that the health of the public and the environment will be enhanced if all of the stakeholders have arrived at a general consensus.

The decision on the level of technology to be employed is not simply one of cost. Decision makers need to be careful not to limit future options to upgrade the system as standards of living increase and demands for level of service increase as well. It would be wasteful to invest in a level of technology only to see it abandoned after a short lifetime. Immediate choices should not preclude future adoption of new more efficient technology.

The Transition

NIS face the difficult but essential task of upgrading water supply and sanitation services. The consumers will have to adjust to a regime in which they pay a greater share of the costs of providing water service without regard to whether the service is provided by the public or private sector. In the interim some of the limited development aid or grants available may need to be used to facilitate that transition in the form of temporary subsidies to those least able to afford full costs of service. Further, there may be a need for demonstration or pilot-scale projects supported by International Financial Institutions to document the benefits of creative public-private partnerships in the water service area.

Conclusions

In the simplest of terms, most water operators want to run a profitable service in a politically stable environment, one that has a sound regulatory, economic and legal system in place. If any of these basic ingredients are missing, then the risk of doing business goes up. The potential operator will then do one of two things. Either the operator will not offer his services and will look elsewhere for more a more favourable climate for doing business. Alternatively, the operator will charge a risk premium, which, in its judgement, provides an adequate incentive to accept the risks as that company perceives them. These risk premiums impose higher costs than those charged in countries with a more favourable business climate, and it is unlikely that the NIS could afford this situation.

Bankers or other financial institutions, which provide capital for investment, should be considered as a key partner in any prospective deal. If international lending institutions like EBRD or the World Bank participate in the project, they should be involved in the earliest stages of the negotiation process.

Early involvement of all the key stakeholders is a crucial element which, while adding cost and complexity, can pay large dividends later on. In particular the customers and the labour force must be consulted.

Economic, social and legal framework conditions are the core considerations that any responsible business partner will want to assess and evaluate before entering into any formal negotiation process.

The capacity of the government authorities to regulate in a fair and reasonable manner is another primary consideration.

Notes

1. World Water Vision – Commission Report – A Water Secure World – April 2000
2. The OECD document on private sector participation prepared for this conference makes these points in a coherent manner.

Chapter 7

POSITION PAPER OF THE NON-GOVERNMENTAL ORGANISATIONS ON THE MAIN DISCUSSION ISSUES OF THE ALMATY MINISTERIAL CONSULTATIONS

We, the Non-Governmental Organisations (NGOs) from the New Independent States (NIS), welcome the Ministerial Consultations in Almaty, Kazakhstan, 16-17 October 2000. The issues proposed for the agenda of the Consultations are of great importance for the NIS region and cover the areas where national governments of the NIS need to pursue active reforms.

We broadly support the main positions, presented in meeting documents: the integration of environmental and economic decision-making, the finance strategies, the reform of urban water supply and sanitation sector and the private sector participation. At the same time, we consider it important to attract the attention of the participants of the Almaty Consultations to the following issues:

1. **Integrating economic and environmental decision-making**

    1.1. We believe that the discussion of this agenda item by the Ministers of Economy/Finance and Environment will be extremely important. In order to achieve significant progress in this area, **it is essential to promote greater awareness in society about the importance of the integration.** To this end, it would be important to organise information campaigns to draw attention of decision-makers at all levels, and of the broad public, to the links between environmental conditions, public health and the rate and quality of economic development at micro and macro levels.

    1.2. Taking into account the specific aspects and the uniqueness of the transition to a market economy, we suggest that the Ministries of economy/finance and environment, with the assistance of international organisations and donors, should undertake the following steps:

    - To review the official methodologies for valuing environmental damages and assessing natural resources currently in use in the NIS, and to adjust them to the new, emerging conditions of market economies; non-governmental organisation will need to be involved in this activity.

    - To organise a series of seminars on valuation of costs and benefits for experts, consultants and governmental officials working in environmental and resource-management departments of the national governments.

    - To provide regular training for public officials on the main elements of, and procedures for, cost-benefit analysis including environmental and nature-management issues.

    - To organise a series of training seminars on the mechanisms for public involvement in decision-making in areas with potential environmental impact for officials in the Ministries of economy/finance and environment.

2. Reform of the urban water supply and sanitation sector

2.1. During the process of designing and implementing reforms of the urban water sector in the NIS, the governments should remember that the *final goal of the reforms is the improvement of the wellbeing of the population, protection of public health and the environment.*

We believe that implementation of the "Guiding principles for reform of the urban water supply and sanitation sector in the NIS" should be oriented towards *gradual integration of urban water management into water basin management, as well as in the territorial management system overall.*

2.2. When determining the general direction of water sector reforms, **strategic environmental impact assessment (SEIA)** should be carried out to analyse alternative approaches and their social, economic and environmental impacts. Public should be actively involved in the SEIA.

2.3. National governments need to be aware of *the role of the "Guiding principles" in the overall process of water sector reforms in the NIS.* It should be clarified that this document applies exclusively to the urban areas with the central water supply and sanitation systems. Such clarification is needed since, in a number of NIS, urban and rural water management are combined in a single framework.

2.4. National governments should *elaborate national plans (programs) for the reform of the urban water supply and sanitation sector,* which would identify main areas of reforms and specific measures, their time lines (long, medium and short-term perspective), as well as sequencing and resource availability.

2.5. *The public should be actively involved* in the reform of the urban water sector. We support the authors of the "Guiding principles", who reflected the importance of public participation in the reform process in a separate section.

We would like to stress the following principles, which need to be followed by national governments, local authorities and water utilities:

- The public authorities and vodokanals should actively provide information on the urban water reform to the public and to NGOs from the outset of the process. A system of information registers should be established to facilitate active and addressed provision of information to target groups of the population.

- Democratically elected public representatives should participate in all working groups, councils and commissions involved in the urban water sector reform.

- Procedures for public consultations and public oversight need to be developed for all stages of the reform (public opinion polls, public hearings, open sessions to respond to public complaints, joint inspections); the public should be involved in the scrutiny of public expenditures in the urban water sector, including the provision of support to socially vulnerable groups of the population.

- The public should participate in the elaboration and conclusion of contracts between the population and the vodokanals, and/or in procedures ensuring that the vodokanal complies with the provisions set out in its contract with the local authority. Prior to the conclusion of a contract with a vodokanal, local authorities need to establish a procedure guaranteeing compliance by both parties, as well as a mechanism for the protection of the rights of the population. The local authorities should appoint their staff specially to monitor the compliance with contracts and to respond to public complaints.

- Support should be provided to initiatives which aim to strengthen co-operation between national and local authorities, vodokanals and the public in the areas of information dissemination, studies of public opinion, demand for services provided by vodokanals and willingness to pay by the population.

- Regular public participation in "state environmental examinations", and the possibility of public environmental examination, should be ensured. The potential and capacity of NGOs should be used effectively for the resolution of conflicts and for representing consumer interests in courts.

- Information education and awareness-raising programmes for the population should be elaborated using the NGO potential.

2.6. **We consider it impossible to raise tariffs for water supply and sewage before the preliminary conditions for the creation of market institutional environment are met,** which include:

- Establishing a clear regulatory framework in this area, including legal provisions for the modification of tariff formula.
- Establishing provisions which allow factors such as willingness and ability to pay for water, and water quality and saving potential, to be taken into account in the tariff setting procedure.
- Establishing a guarantee mechanisms to ensure that vodokanals fulfil their obligations to the clients (population, industrial enterprises, budget organisations).
- Ensuring the transparency of tariff setting procedures for consumers.
- Wide-scale installation of water meters.
- Establishing a mechanism for targeted support to the poor and careful consideration of a one-off amnesty of debts accumulated by the poor.

2.7. We welcome the efforts to identify additional sources of financing for the operations of vodokanals. Private sector participation in the development of the water sector can improve the operation of vodokanals and provide financing for their development. However, international experience provides several negative examples in this field. Therefore, **the need for private sector participation, as well as its specific forms, should be carefully analysed in each case and with obligatory public participation.** Investment tenders should be based on clear rules, established in consultation with all stakeholders, reflecting stakeholders' interests to the maximum possible degree; tender procedures should be open for public participation. Public authorities and the private sector should strictly comply with their obligations. Legislative authorities should pay special attention to the need to adopt legal acts with direct effect, which would facilitate the inflow of investments into the water sector. The judiciary system should be strengthened, as well as the enforcement of legal acts regulating the water and other sector.

We suggest to the participants to elaborate a Program for public involvement in the implementation of the decisions of the Almaty Ministerial Consultations with the participation of the European EcoForum, other interested NGOs and the New RECs from the NIS.

This Position Paper has been developed on the basis of consultations organised by the EAP Task Force Group of the European Eco-Forum, which involved 111 non-governmental organisations from all NIS. We would like to extend out gratitude for the assistance provided for these consultations by the Danish EPA, Finnish Ministry of environment, Ministry of environment of Norway, the EAP Task Force Secretariat and the Regional Environmental Centre of the Russian Federation.

Annex 1

GUIDING PRINCIPLES FOR REFORM OF THE URBAN WATER SUPPLY AND SANITATION SECTOR IN THE NIS

Introduction

Several recent studies have documented that the water supply and sanitation services in the New Independent States of the former Soviet Union (NIS) are in critical condition and deteriorating.[1] There are frequent interruptions in service, yet water consumption is excessively high by international standards with considerable wastage. Water quality is deteriorating, and in some NIS resulting in adverse impacts on human health, productivity and important ecosystems.

Although investments are required, these problems will not be solved only by increasing the supply of financial resources from public budgets: such an approach would not be sustainable and would reinforce the inefficiency of existing arrangements. In any case, the financial resources required to address the scale of the present problem simply do not exist. In some countries only 30-40% of the resources needed to operate and maintain the existing networks are available, and external financial resources could only help to address a small fraction of the total needs.

Important initiatives to reform vodokanals (water utilities) have been launched at the local level in some countries which are yielding valuable experience, and they should be continued. However, it is questionable whether these initiatives can be widely replicated unless new policies are implemented establishing a sounder, more sustainable basis for sector reform. Some local initiatives have involved support from donors and/or International Financial Institutions (IFIs); but, overall, the results achieved from these projects have been mixed, with the results of some projects disappointing compared with the effort invested. This generates frustration for both local and foreign partners.

Experience gained since 1991 clearly indicates that a new framework is needed to guide the reform of the urban water supply and sanitation sector. Such a framework is essential to stop the continued deterioration and eventual collapse of water and sanitation services, with the serious consequences for the health of the population and their environments which this would entail. The objective of the reform should be to address these problems and to ensure that good quality water and sanitation services are delivered reliably, sustainably and at least cost to the population. Some of the key reforms needed to achieve this objective include:

- Decentralising responsibility for water service provision from national authorities to the local level.
- Reforming vodokanals so that they have the autonomy, capacity and means to provide water and sanitation services efficiently and effectively, based on a realistic assessment of needs and subject to strict supervision by public authorities.
- Engaging the public directly in the reform process.
- Establishing the sector on a financially sustainable basis so that funds are available to cover operation and maintenance costs and to make necessary investments, while addressing the needs of poor and vulnerable households.
- Creating incentives to substantially increase efficiency in the use of water by consumers and in the operation of vodokanals.

The Guiding Principles set out below are intended to present a "vision" of the types of reform which are needed. They suggest how the roles and responsibilities of different stakeholders should be changed in order to ensure that safe water and high quality sanitation is provided reliably and cost-effectively. They should be seen as an inter-related group of recommendations where new rights (e.g. for vodokanals) are balanced by new responsibilities. The Guiding Principles address the urban water sector; they would need to be adapted to apply to rural water supply. More generally, urban and rural water reform should form part of integrated river basin management systems.

The Guiding Principles are not strict "rules"; rather they propose a direction for reform. They reflect "best practices" which have been successfully applied in other parts of the world, including in Central and Eastern Europe. They have been developed through a participatory process involving major stakeholders. NIS governments, donors, IFIs, NGOs, the private sector and other stakeholders are invited to take the Guiding Principles into account when reforming the urban water supply and sanitation sectors.

The Guiding Principles clearly need to be elaborated further and adapted to the particular circumstances in each NIS. This process of adaptation will need to be integrated into the broader process of reform, and to take account of the important differences in the financial capacities and structures of the urban water supply and sanitation sector of the various NIS. However, the NIS also share many common features which could underpin a process of co-operation and experience sharing, implemented through the EAP Task Force.

The Guiding Principles are set out immediately below:

I. Key Elements of Urban Water Sector Reform

1. *Establishing Strategic Objectives*

The main strategic objective of urban water sector reform should be to ensure that good quality water and sanitation services are delivered reliably, sustainably and at least cost to the population. This objective should be translated into national action programmes which contain realistic, quantitative, time-bound targets covering performance and quality objectives.

The national authorities should facilitate participatory processes involving governmental and non-governmental stakeholders to establish targets, and to develop a strategy for achieving them, which prioritises actions to minimise adverse impacts on human health, productivity and important ecological systems. Urban water sector reform strategies should ultimately be integrated into river basin management schemes.

Finance strategies should be developed in which the feasibility of targets are assessed against a realistic appraisal of the likely sources and levels of finance; the measures required to ensure that the levels of finance provided from different sources (including vodokanals) are sufficient to achieve agreed targets should be clearly identified.

Socio-economic and willingness-to-pay studies should be carried out to assess the demand for, and affordability of, improved utility services at the level of households; the studies should include an assessment of the needs of poor and vulnerable households.

2. *Reforming Institutions and Clarifying Their Roles*

The Role of the National Authorities should be to set the framework for managing urban water supply and sanitation by:

Decentralisation:

- Decentralising responsibility for water supply and sanitation services to the municipal level, avoiding excessive fragmentation.
- Destablishing the legal, regulatory and institutional framework for sound and sustainable municipal finance, inluding effective planning, supervision and fiscal control systems for municipalities.
- Clarifying the legal status of vodokanals, their relations with local governments and property rights for infrastructure.
- Establishing a framework for treating the inherited debts of vodokanals.

Regulatory Oversight:

- Depending on the particular circumstances in a country, consider establishing an independent, national regulatory agency to ensure that vodokanals do not exploit a monopoly position and/or to protect them from undue political interference. In such cases, the objectives of the regulation should be clearly identified and appropriate means for achieving them provided.
- Regulating issues that have national or inter-municipal dimension, such as standards for environmental quality, wastewater discharge and drinking water; and establishing the legal framework to facilitate water and sanitation management initiatives undertaken jointly by groups of municipalities.
- Establishing the legal and regulatory framework for stakeholder involvement, including private sector participation and consumer protection.
- Establishing a framework for managing the competitive uses of water at the national and regional levels, including principles and rules for the management of different water resources, and policies for integrating municipal water and sanitation systems into coherent programs for water resources management within river basins.
- Ensuring that an adequate system for monitoring water quality is in place and that the results are available to the public.

Strategy Formulation and Technical Assistance

- Defining strategic policies and development objectives, including investment strategies and the means for financing them; such policies and investment strategies should strike an appropriate balance between water supply and sanitation objectives.
- Providing assistance to utilities and local governments in areas such as capacity building, finance, and international assistance co-ordination.
- Promoting demonstration projects to reform selected vodokanals; disseminating results; publishing performance indicators for vodokanals.
- Facilitating market creation and promoting competition in the supply of goods and services to vodokanals.

The Roles of Local Government[2] and Vodokanals should be clarified. Local governments should establish vodokanals as autonomous, commercially-run utilities, with responsibility for billing and collecting payments. They also need to build capacity to exercise effective regulatory oversight over utilities. This would involve:

Transparent Transfer of Responsibilities to Vodokanals:

- Over time, the responsibility for setting and adjusting tariffs should be vested with vodokanals, subject to the supervision of local government or an independent regulatory agency. The public should also be involved in the process.
- The right of vodokanals to use assets over long periods of time should be established unambiguously. When ownership of assets is to be retained in the public sector, the assets should be transferred to the lowest level of government; this should not be lower than that of the service provider.
- Local and national governments should develop a plan for dealing with vodokanals' debt and removing obstacles to their raising capital for investment; options for payables include reorganisation and debt relief from the national authorities or liquidation; options for receivables may involve a once-only subsidy.

Regulatory Oversight:

- Local government should hold vodokanals accountable for their performance, *e.g.* using performance contracts which clearly define performance and quality targets to be achieved by the vodokanal and the support to be provided by local government. The contracts should include appropriate incentives and sanctions to encourage efficient service provision by vodokanals, as well as a requirement to issue regular reports to the public on progress in achieving performance targets.
- Local governments should subject vodokanals to hard budget constraints and not provide support, in any form, in addition to that agreed in performance contracts.
- In the transition period (and subsequently), vodokanals should be required to prepare proposals, supported by detailed justification, for tariff rates and structure, taking into account the recommendations on tariffs described above.

Strategy Formulation and Private Sector Participation

- When developing a strategy for water sector reform, local governments and vodokanals should consider inviting private operators to participate in providing water and sanitation services. The arrangement selected should be consistent with the government's reform strategy and capacity.[3]
- When contracting a private utility operator, interested governments should retain the assistance of experienced legal, technical and financial advisors with substantial international experience. Decisions to invite private sector participation should be based on competitive bidding procedures. The process should be open and transparent, with adequate provision for public participation.
- Vodokanals should develop and implement a program of comprehensive investment planning, covering operation, maintenance, rehabilitation and eventually capital investments, based on realistic estimates of future revenues.

The Public should be directly engaged in the reform process. Reform of the urban water sector will change the role of the public from essentially passive consumers of state-provided services, to purchasers of those services. This will require greater involvement of the public in decisions concerning the level and type of service provision, and the associated financial implications. Government and vodokanals should provide for effective public decision-making in sector reform and ensure that poor and vulnerable groups have adequate access to water services:

- National authorities should establish a legislative basis for public participation in key decisions concerning water supply and sanitation services.

- Local government and vodokanals should actively provide consumers with full information and opportunities to participate in key decisions concerning water supply and sanitation, especially through public meetings and participation in decision-making bodies.
- Performance contracts between local governments and vodokanals should be developed through a participatory process and ensure that the interests of all stakeholders, particularly the poorest and most vulnerable, are protected.
- Government, and not vodokanals, should have the ultimate responsibility for ensuring that poor and vulnerable households have adequate access to water and sanitation services; transparent, targeted and efficient subsidies which take account of tariffs for all utilities and address integrated household needs should be used to provide support to such households.

3. *Establishing a Framework for Financial Sustainability*

The following measures are important to create conditions for financial sustainability of vodokanals:

- Improve coverage and efficiency of billing and payment collection so as to maximise revenues; introduce and enforce sanctions for non-payment, including cut-offs as the last resort.
- Phase out as quickly as possible non-cash forms of payment, where such practices exist.
- Settle outstanding payments by public organisations to the vodokanal; make these organisations responsible for ongoing payments for water services; if financial support is needed, it should be through transparent government subsidies.
- Replace the current "basic cost-plus" tariff formula with one that provides incentives for cost reductions and allows for an acceptable level of profits.
- Reduce large differentials in tariffs between household, industrial, and other users; this will likely encourage more rational water use by those who are currently cross-subsidised, and reduce resistance to pay their water bill by those who currently pay very high tariffs.
- Establish a procedure for periodic tariff revision and adjustment, which is transparent, involves the participation of stakeholders, and is indexed to inflation.
- Develop a program for gradual tariff increases to cover operation, maintenance, rehabilitation and ultimately investment costs; such increases should take full account of affordability constraints and be part of a strategy for service improvement which has been developed through a participatory process.
- Introduce accounting systems that conform with international standards; national authorities should determine tax liability on this basis.

4. *Promoting Efficiency and Cost-Effective Use of Resources*

Governments should establish an operational framework for vodokanals which promotes the efficient use of water and other resources; this framework should encourage vodokanals to develop new approaches for promoting the efficiency improvements in networks and plants which are urgently needed.

National standards which are stricter and more costly than those established internationally should be reviewed and modified so as to reduce compliance costs to a realistic level.

A new management culture needs to be introduced to water sector institutions based on clear responsibilities, incentives for initiative and good performance, and accountability; this should be supported by appropriate management and salary structures, performance-related pay and computerised information systems.

Human resources management policies should promote appropriate staff training and, in consultation with the work force, the achievement of efficient staff levels. Areas where capacity building is urgently required include: business management, economics, law, consumer awareness, needs assessment techniques and processes for facilitating dialogue and co-operation between vodokanals and the public. The revision of engineering curricula of universities also requires urgent attention.

A metering strategy should be designed and implemented to reduce consumption, wastage and operational costs and to improve revenue collection; the efficiency and transparency of metering apartment blocks should be improved as a step toward the ultimate goal of metering individual users; investments in meters should be commensurate with benefits.

Vodokanals should develop public outreach programmes which encourage the more efficient use of water by consumers as an integral part of efforts to improve the quality and reliability of water services.

Limited investment resources should be concentrated on projects:

- To reduce operating costs through high pay - back investments e.g. demand management, leak reduction and energy savings.
- To improve the safety and reliability of water supply services, particularly to address public health concerns.
- To maintain and rehabilitate essential elements of wastewater collection and treatment infrastructure, particularly to address public health concerns.

II. Sequencing Reforms

Strategies to reform the water supply and sanitation sector should sequence actions and prioritise those on which other steps depend. Such strategies need to be integrated with, and take full account of, the broader process of economic, political and social reform. Since some actions will require legislative or other time-consuming processes, they should be initiated as soon as possible, or accelerated if already in progress. Priority actions should include:

- Launching participatory, multi-stakeholder processes to support the development and implementation of the strategies.
- Decentralising authority to the local level.
- Creating autonomous vodokanals and, as required, an independent regulatory agency.
- Establishing performance contracts between local governments and vodokanals, initially with a focus on improving service levels through affordable, low-cost measures.
- Strengthening the financial stability of vodokanals by reforming tariff levels, collection systems and debt.

Notes

1. UK Department for International Development (2000), "Obstacles and Opportunities to Commercialising Urban Water, Services in the New Independent States"; Danish Ministry of Energy and Environment (2000) "Environmental Finance Strategies for Georgia, Moldova and Novgorod"; World Bank (2000) "Water Supply and Sanitation Sector Review and Strategy for Azerbaijan". Some of the main findings of these studies are summarised in the Background Document on this issue prepared for the Almaty Ministerial Consultation CCNM/ENV/EAP/MIN(2000)5.
2. "Local government" is used to mean the appropriate level of sub-national government, whether at the regional or municipal level.
3. The options and issues associated with private sector participation in the urban water sector are analysed in the background document on this subject prepared for the Almaty Conference CCNM/ENV/EAP/MIN(2000)7.

Annex 2

LIST OF PARTICIPANTS

ARMENIA	**Mr. Murad MURADYAN** Minister of the Environment 35, Moskovyan Str. 375002 Yerevan	Tel: +3741 59 10 99 / 53 18 61 Fax: +3741 53 18 61 E-mail: nuneemil@yahoo.com, interdpt@freenet.am
	Mr. Merujan MIKAELYAN Deputy Minister of Finance and Economics 1, Melik Adamyan Str. 375010 Yerevan	Tel: +3741 59 53 08 / 50 93 09 Fax: +3741 15 10 69 E-mail: hasmikpress@yahoo.com
	Mr. Ashot ARUTYUNYAN Head of Economic Department Ministry of the Environment 35, Moskovyan Str. 375002 Yerevan	Tel: +3741 59 10 99 / 53 18 61 Fax: +3741 53 18 61 E-mail: nuneemil@yahoo.com, interdpt@freenet.am
	Mr. Volodia NARIMANYAN Head of the Water Management Ministry of the Environment 35, Moskovyan Str. 375002 Yerevan	Tel: +3741 59 10 99 / 53 18 61 Fax: +3741 53 18 61 E-mail: nuneemil@yahoo.com, interdpt@freenet.am
AZERBAIJAN	**Mr. Nadim KAZIBEKOV** First Deputy Chairman Committee for Melioration and Water Management The House of Government, 370016 Baku	Tel: +994 12 93 61 54 Fax: +994 12 93 11 76
	Mr. Farukh DADASHEV Head of Department for Capital Investments Ministry of Economy The House of Government 370016 Baku	Tel: +994 12 93 81 60 Fax: +994 12 93 20 25
	Mr. Mammed ASADOV Head of Depart. "Science, Project and Expertise" Committee for Amelioration and Water Management The House of Government 370016 Baku	Tel: +994 12 93 61 54 / 93 80 11 Fax: +994 12 93 11 76
BELARUS	**Mr. Mikhail RUSY** Minister of Natural Resources and Environment 10, Kollektornaya Str. 220048 Minsk	Tel: +375 17 220 66 91 Fax: +375 17 220 55 83 / 220 47 71 E-mail: minproos@minproos. belpak.minsk.by
	Ms. Nadezhda URUPINA Head of Department Ministry of Finance 7, Sovyetskaya Str. 220048 Minsk	Tel: +375 17 222 61 37 Fax: +375 17 220 21 72

	Mr. Alexander RACHEVSKY Head of International Co-operation Department Ministry of Natural Resources and and Environment 10, Kollektornaya Str. 220048 Minsk	Tel: +375 17 220 43 28 Fax: +375 17 220 55 83 / 220 47 71 E-mail: minproos@minproos. belpak.minsk.by
CZECH REPUBLIC	Mr. Jirí HLAVÁCEK Deputy Minister – Director General International Relations Section Ministry of the Environment Vrsovicka 65 100 10 Praha 10	Tel: +420 26712 2916 Fax: +420 26731 0307 E-mail: jiri_hlavacek@env.cz
	Ms. Helena CÍZKOVÁ Adviser to Deputy Minister International Relations Section Ministry of the Environment Prokešovo nari. 8 702 00 Ostrava	Tel: +420 69628 2362 Fax: +420 69611 8798 E-mail: cizkova@env.cz
DENMARK	Mr. Svend AUKEN Minister of Environment and Energy Hojbro Plads 4 DK-1200 Copenhagen K.	Tel: +45 3392 7600 Fax: +45 3332 2227 E-mail: mem@mem.dk
	Mr. Leo BJORNSKOV Under-Secretary of State for International Relations Ministry of Environment and Energy Hojbro Plads 4 DK-1200 Copenhagen K.	Tel: +45 3392 7600 Fax: +45 3332 2227 E-mail: mem@mem.dk
	Mr. Karsten SKOV Deputy Director General, DEPA Ministry of Environment and Energy Strandgade 29 DK-1401 Copenhagen K.	Tel: +45 3266 0100 Fax: +45 3266 0479 E-mail: kas@mst.dk
	Mr. Palle LINDGAARD-JORGENSEN Head of Department DEPA Ministry of Environment and Energy Strandgade 29 DK-1401 Copenhagen K.	Tel: +45 3266 0100 Fax: +45 3266 0479 E-mail: plj@mst.dk
	Mr. Soren BUCH SVENNINGSEN Special Advisor to the Minister Ministry of Environment and Energy Hojbro Plads 4 DK-1200 Copenhagen K.	Tel: +45 3392 7600 Fax: +45 3332 2227 E-mail: mem@mem.dk
	Mr. Erik TANG Project Coordinator DEPA Ministry of Environment and Energy Strandgade 29 DK-1401 Copenhagen K.	Tel: +45 3266 0100 Fax: +45 3266 0479 E-mail: eta@mst.dk
FINLAND	Mr. Jaakko HENTTONEN Environment Counsellor Central and East European Cooperation Ministry of the Environment Kasarmikatu 25, P.O. Box 380 00131 Helsinki	Tel: +358 400 165 805 Fax: +358 9 1991 9515 E-mail: jaakko.henttonen@vyh.fi

FRANCE	**Mr. Alain RENOUX** Director's Assistant General Directorate for Administration of Financial and International Business Ministry of Spatial Planning and Environment 20, Segur Ave. 75302 Paris 07	Tel: +33 1 421 91 636 Fax: +33 1 421 91 832 E-mail: alain.renoux@ environnement.gouv.fr
	Mr. Jacques SIRONNEAU Head of Legal Office Water Department Ministry of Environment 20, Segur Ave. 75302 Paris 07	Tel: +331 421 91 270 Fax: +331 421 91 269 E-mail: jacques.sironneau@ environnement.gouv.fr
GEORGIA	**Ms. Nino CHKHOBADZE** Minister of Environment and Natural Resources 68a, Kostava Str. 380015 Tbilisi	Tel: +995 32 23 06 64 Fax: +995 32 33 3952 / 94 36 70 E-mail: gmep@caucasus.net
	Ms. Natia TURNAVA Deputy Minister of Economics, Industry and Trade 12, Chanturia Str. Tbilisi	Tel: +995 77424441 Fax: +995 32 93 33 13
	Ms. Neli KHIDESHELI Head of Division Ministry of Economics 12, Chanturia Str. Tbilisi	Tel: +995 7742 4441 Fax: +995 32 9333 13
	Mr. Zaal LOMTADZE Head, Department of Environmental Policy Ministry of Environment 68a, Kostava Str. 380015 Tbilisi	Tel: +995 32 23 06 64 Fax: +995 32 33 39 52 / 94 36 70 E-mail: zl@gol.ge
	Mr. Malkhaz ADEISHVILI Deputy Head Department of Environmental Economics Ministry of Environment 68a, Kostava Str. 380015 Tbilisi	Tel: +995 32 33 22 43 / 23 06 64 Fax: +995 32 33 39 52 E-mail: envecon@caucasus.net
GERMANY	**Ms. Gila ALTMANN** State Secretary Ministry of Environment Alexanderplatz 6 10178 Berlin	Tel: +49 1888 305 2040 Fax: +49 1888 305 4375 E-mail: altmann.gila@bmu.de
	Mr. Jürgen KEINHORST Head of Division International Affairs, Co-operation CEE/NIS Ministry of Environment Alexanderplatz 6 10178 Berlin	Tel: +49 1888 305 2370 Fax: +49 1888 305 4375 E-mail: Keinhorst.Juergen@bmu.de
	Mr. Heinrich SCHMAUDER Specialist International Affairs and Co-operation Dept. Federal Agency for Nature Conservation Konstantinstr. 110 53179 Bonn	Tel: +49 228 8491 241 Fax: +49 228 8491 245 E-mail: schmauder_h@bfn.de
	Mr. Zeno REICHENBECHER International Environmental Protection Dept. Ministry of Economics and Technology Scharnhorst Str. 34-37 10115 Berlin	Tel: +49 30 2014 7115 Fax: +49 30 2014 5428 E-mail: reichenbecher@ bmwi.bond.de

	Mr. Anja VON ROSENSTIEL Programme Officer International Affairs, Co-operation CEE/NIS Ministry of Environment Alexanderplatz 6 10115 Berlin	Tel: +49 1888 305 2378 Fax: +49 1888 305 4375 E-mail: rosenstiel.anja@bmu.de
	Ms. Gabriela SCHEUFELE Consultant at the CCD project (GTZ) Deutsche Gesellschaft fur Technische Zusammenarbeit Wachsbleiche, 1 53111 Bonn	Tel: +49 228 98 37 112 Fax: +49 228 98 37 125 E-mail: Scheufele@gtzccd.de
KAZAKHSTAN	**Mr. Serikbek DAUKEYEV** Minister of Natural Resources and Environmental Protection 1, Satpaev Str. 475000 Kokshetau	Tel: +7 31622 54265 Fax: +7 31622 50620 E-mail: dnasir@koksh.kz
	Mr. Mazhit ESENBAEV Minister of Finance	
	Mr. Zhaksybek KULEKEYEV Minister of Economics The House of Ministries Astana	Tel: +7 3172 117511 / 117512 Fax: +7 3172 118157 E-mail: mineconom@nursat.kz
	Mr. Tito SYZDYKOV Chairman of the Environmental Committee of Parliament	
	Mr. Viktor KHRAPUNOV Major of Almaty city	
	Mr. Amanbek RAMAZANOV Chairman of the Water Resources Committee	
	Mr. B. S. ABDREEV Head of Management Department for Sectoral Policy Ministry of Economy	
	Mr. Vasiliy ZUEV General Director of Gorvodokanal of Almaty city	
	Mr. Serikbai NURGISAEV Governor of Kyzylorda region	
	Mr. Murat MUSATAYEV Vice Minister Ministry of Natural Resources and Environmental Protection 1, Satpaev Str. 475000 Kokshetau	Tel: +7 31622 542 69 Fax: +7 31622 506 20 E-mail: mmusataev@neapsd.kz
	Mr. Kairat AITIKENOV Chairman of Environment Protection Committee 1, Satpaeva Ave. 475000 Kokshetau	Tel: +7 31622 554 10 Fax: +7 31622 554 31 E-mail: aabdraghmanova@koksh.kz
REPUBLIC OF KYRGYZSTAN	**Mr. Tynybek ALYKULOV** Minister of Environmental Protection 131, Isanov Str. 720033 Bishkek	Tel: +996 312 21 48 45 Fax: +996 312 21 36 05 E-mail: min-eco@elcat.kg
	Mr. Muktarbek SULAIMANOV Head of Department for Foreign Relations Ministry of Environment 131, Isanov Str. Bishkek	Tel: +996 312 21 48 45 Fax: +996 312 21 36 05 E-mail: min-eco@elcat.kg

Annex 2

                    Mr. Bulat AZYKOV   Tel: +996 312 66 35 31
Head of Department   Fax: +996 312 66 19 55
Ministry of Finance   E-mail: aziz27@sti.gov.kg
58, Erkindik Str.
720040 Bishkek

REPUBLIC OF MOLDOVA   **Mr. Alexander JOLONDCOVSKIY**   Tel: +373 2 228 608
First Deputy Minister   Fax: +373 2 220 748
Ministry of Environment and Territorial   E-mail:
Development   relint@mediu.moldova.md
9, Cosmonautilor Str.
MD-2005 Chisinau

Mr. Viktor ZUBAREV   Tel: +373 2 234 056
Head of Main Dept. for Natural Resources   Fax: +373 2 220 748
Ministry of Economy and Reforms   E-mail: victor_zubarev@
1, Piata Marii Adunari Nationale Str.   hotmail.com
MD-2001

Mr. Valeriu BINZARI   Tel: +373 2 228 608
Head of Department   Fax: +373 2 220 748
Main Department for Capital Investments   E-mail:
Ministry of Finance   relint@mediu.moldova.md
7, Cosmonautilor Str.
MD-2005 Chisinau

Mr. Andrei ISAC   Tel: +373 2 226 254
Head of Environmental Policy Division   Fax: +373 2 220 748
Ministry of Environment and Territorial   E-mail: a.isac@moldova.md
Development
9, Cosmonautilor Str.
MD-2005 Chisinau

THE NETHERLANDS   **Mr. Hugo VON MEIJENFELDT**   Tel: +31 70 339 47 19
Deputy Director   Fax: +31 70 339 13 06
Directorate for International Environmental   E-mail: hugo.vonmeijenfeldt@
Affairs   minvrom.nl
Ministry of Housing, Spatial Planning
and the Environment
P.O. Box 30945, IPC 670
2500 GX The Hague

Mr. Reginald HERNAUS   Tel: +31 70 339 46 79
Senior Policy Advisor   Fax: +31 70 339 13 06
Directorate for International Environmental   E-mail: reggie.hernaus@
Affairs   minvrom.nl
Ministry of Housing, Spatial Planning
and the Environment
P.O. Box 30945, IPC 670
2500 GX The Hague

Ms. Marion FOKKE-BAGGEN   Tel: +31 70 339 49 54
Water Department   Fax: +31 70 339 12 88
Ministry of Housing, Spatial Planning   E-mail: Marion.FokkeBaggen@
and the Environment   minvrom.nl
P.O. Box 30945
2500 GX, The Hague

NORWAY   **Ms. Eldrid NORDBO**   Tel: +47 222 45 980
Director General of the Department for   Fax: +47 222 42 755
International Co-operation,   E-mail: eno@md.dep.no
Climate and Polar Affairs
Ministry of the Environment
Myntgt. 2, P.O. Box 8013
Oslo – Dep., N-0030 Oslo

	Mr. Richard FORT Senior Adviser Department for International Co-operation, Climate and Polar Affairs Ministry of the Environment P.O. Box 8013 Oslo – Dep., N-0030 Oslo	Tel: +47 222 45 982 Fax: +47 222 49 564 E-mail: rgf@md.dep.no
RUSSIAN FEDERATION	Mr. Mukhamed TSIKANOV Deputy Minister Ministry of Economic Development and Trade 1-3, First Tverskaya-Yamskaya Str., A-47 125818 Moscow	Tel: +7 095 250 13 33 Fax: +7 095 251 69 65 E-mail: dito@gov.ru
	Mr. Alexey PORYADIN First Deputy Minister Ministry of Natural Resources 4/6, Bolshaya Gruzinskaya Str. 123812 Moscow	Tel: +7 095 254 50 77 Fax: +7 095 943 00 13 / 254 46 10
	Mr. Evgeniy SHOPKHOEV Head of Department for Economics of Nature Use, Environmental Protection and Programmers for Emergency Situations Ministry of Economic Development and Trade 1-3, First Tverskaya-Yamskaya Str., A-47 125818 Moscow	Tel: +7 095 200 04 28 / 209 83 94 Fax: +7 095 209 80 58 E-mail: dito@polinom.ru
	Ms. Nina DOBRYNINA Chief Specialist Ministry of Natural Resources RF 4/6, Bolschaya Gruzinskaya Str. 123812 Moscow	Tel: +7 095 230 87 22 Fax: +7 095 943 00 13
	Mr. Pavel KOSYKH Secretary of EcoCouncil Parliamentary Union	
SWITZERLAND	Mr. Jürg SCHNEIDER Senior Program Officer for Development Co-operation, International Affairs Division Agency for the Environment, Forests and Landscape CH-3003 Berne	Tel: +41 31 322 68 95 Fax: +41 31 323 03 49 E-mail: juerg.schneider@ buwal.admin.ch
TADJIKISTAN	Mr. Usmonkul SHOKIROV Minister of Nature Protection 12, Bokhtar Str. 734025 Dushanbe	Tel: +992 372 21 30 39 Fax: +992 372 21 30 39 / 21 18 39 E-mail: shokirov@tojikiston.com
	Mr. Viktor BOLTOV Vice-Minister of Economy and Foreign Economic Relations 42, Rudaki Ave. 734025 Dushanbe	Tel: +992 372 21 64 00 Fax: +992 372 21 69 14 E-mail: vboltov@tajnet.com
	Mr. Munimjon ABDUSAMATOV Head of Water Inspection Ministry of Nature Protection 12, Bokhtar Str. 734025 Dushanbe	Tel: +992 372 21 56 69 Fax: +992 372 21 30 39 / 21 18 39 E-mail: stihiya@tajnet.com, shokirov@tojikiston.com
	Ms. Tatiana ALIKHANOVA Chief Specialist Ministry of Economy and Foreign Economic Relations 42, Rudaki Ave. 734025 Dushanbe	Tel: +992 372 21 30 80 Fax: +992 372 21 69 14

Annex 2

TURKEY	**Ms. Kumru ADANALI** Division Manager Foreign Relations Department Ministry of Environment Eskisehir Yolu 8. Km. 06100 Ankara	Tel: +312 285 17 05 Fax: +312 285 37 39
TURKMENISTAN	**Mr. Begench ATAMURADOV** Deputy Minister of Nature Protection Ministry of Nature Protection 102, Kemine Str. 74400 Ashgabad	Tel: +993 12 398 590 Fax: +993 12 511 613 E-mail: nature@online.tm
	Mr. Makhtumkuli AKMURADOV Head of Department for Information and International Co-operation Ministry of Nature Protection 102, Kemine Str. 74400 Ashgabad	Tel: +993 12 398 594 Fax: +993 12 511 613 E-mail: nature@online.tm
UKRAINE	**Mr. Ivan ZAETZ** Minister of Environment and Natural Resources 5, Khreshchatyk Str. Kiev 0161	Tel: +380 44 228 06 44 Fax: +380 44 229 83 83
	Mr. Iaroslav MOVCHAN Head of Department Department of Protection, Use and Restoration of Natural Resources Ministry of Environment and Natural Resources 5, Khreshchatyk Str. Kiev 0161	Tel: +380 44 228 20 67 Fax: +380 44 228 20 67 E-mail: movchan@ mep.freenet.kiev.ua
	Mr. Vitaliy POTAPOV Head of Dept. for Environmental Policy Ministry of Environment and Natural Resources 5, Khreshchatyk Str. Kiev 0161	Tel: +380 44 228 25 22 Fax: +380 44 228 25 22
UNITED KINGDOM	**Mr. George FOULKES** Parliamentary Undersecretary of State Department for International Development 94, Victoria Str. London SWIE 5JL	Tel: +44 20 7917 0621 Fax: +44 20 7917 0831
	Mr. Rod MATTHEWS Divisional Engineering Adviser Department for International Development 94, Victoria Str. London SWIE 5JL	Tel: +44 20 7917 0552 Fax: +44 20 7917 0072 E-mail: rh-matthews@ dfid.gov.uk
	Ms. Sheila Mary McCABE Divisional Manager of the Env. Protection Intl. Dept. of the Environment, Transport and Regions Ashdown House, 123 Victoria Str. London SWIE 60E	Tel: +44 20 7944 6220 Fax: +44 20 7944 0831 E-mail: sheila_mccabe@ detr.gsi.gov.uk
	Mr. Peter SMITH Environment Engineering and Geoscience Adviser Department for International Development 94, Victoria Str. London SWIE 5JL	Tel: +44 20 7917 0206 Fax: +44 20 7917 0072 E-mail: p-smith@dfid.gov.uk

Mr. Santib BAISYA
Assistant Private Secretary to George Foulks
Department for International Development
94, Victoria Str.
London SW1E 5JL

Tel: +44 20 7917 0182
Fax: +44 20 7917 0634
E-mail: s-baisya@dfid.gov.uk

UNITED STATES

Mr. Gene GEORGE
Director, Office of Environment, Energy
and Social Transition
USAID
1300 Pennsilvania Avenue
Washington, DC 20523

Tel: +1 202 712 12 78
Fax: +1 202 216 34 09
E-mail: Ggeorge@usaid.gov

Mr. Glenn MORRIS
Consultant, Environmental Branch
of Central Europe and Asian Department
USAID
118 Nottingham Drive
Chapel Hill, NC 27514

Tel: +1 919 932 98 13
Fax: +1 919 932 98 13
E-mail:
gemorris@mindspring.com

UZBEKISTAN

Mr. Khalilulla SHERIMBETOV
Acting Chairman
State Committee for Nature Protection
7, Kadiri Str.
700128 Tashkent

Tel: +998 712 41 49 23
Fax: +998 712 41 39 90 /
41 56 33

Mr. Sergey SAMOILOV
Head of Department for Economics
and Management of Nature Use
State Committee for Nature Protection
7, Kadiri Str.
700128 Tashkent

Tel: +998 712 41 51 40
Fax: +998 712 41 39 90 /
41 56 33

Mr. Takhir KARIMOV
Head of Department
Ministry of Finance
5, Mustakkillik Str.
Tashkent

Tel: +998 711 44 53 66
Fax: +998 711 39 16 23

EUROPEAN COMMISSION

Mr. Jean-François VERSTRYNGE
Deputy Director General
Directorate General Environment
200, Rue de la Loi
B-1049 Brussels

Tel: +322 295 11 47
Fax: +322 299 03 10
E-mail: jean-francois.verstrynge
@cec.eu.int

Mr. Norbert JOUSTEN
Head of Unit
Regional Co-operation and Nuclear Safety
200, Rue de la Loi (CHAR 9/40)
B-1049 Brussels

Tel: +322 295 68 71
Fax: +322 296 39 18
E-mail:
Norbert.Jousten@cec.eu.int

Mr. Jaime REYNOLDS
Co-ordinator of Bilateral Cooperation
with NIS
200, Rue de la Loi
B-1049 Brussels

Tel: +322 299 56 32
Fax: +322 299 41 23
E-mail: Jaime.Reynolds@
cec.eu.int

OECD

Ms. Joke WALLER-HUNTER
Director, Environment Directorate
Organization for Economic Cooperation
and Development (OECD)
2, rue André Pascal
75775 Paris Cedex 16
France

Tel: +33 1 45 24 93 00
Fax: +33 1 45 24 78 76
E-mail: joke.waller-hunter@oecd.org

WORLD BANK

Mr. Johannes LINN
Vice-President
Europe and Central Asia
1818 H Str. NW
Washington, DC 20433, USA

Tel: +1 202 458 06 02
Fax: +1 202 522 27 58
E-mail: JLinn@worldbank.org

Mr. Anthony CHOLST
Senior country Officer
Central Asia Country Unit, Europe
and CA region
1818 H Str. NW
Washington, DC 20433, USA

Tel: +1 202 458 03 24
Fax: +1 202 477 33 72
E-mail: acholst@worldbank.org

Mr. Kiyoshi KODERA
Country Director
Central Asia Country Unit, Europe
and CA egion
600 19th Str. NW
Washington, DC 20435, USA

Tel: +1 202 473 35 44
Fax: +1 202 477 33 72
E-mail: Kkodera@worldbank.org

Mr. Kevin CLEAVER
Sector Director
Environmentally and Socially Sustainable
Development
1818 H Street, Room 5-317
Washington, DC 20433, USA

Tel: +1 202 473 45 95
Fax: +1 202 614 06 97
E-mail: Kcleaver@worldbank.org

Ms. Jane HOLT
Sector Manager
Environmentally and Socially Sustainable
Development Dept.
1818 H Str. NW
Washington, DC 20433, USA

Tel: +1 202 458 89 29
Fax: +1 202 614 06 96
E-mail:
Jholt@worldbank.org

Mr. Stefan SCHWAGER
Sr. Environmental Specialist / PPC Officer
Environmentally and Socially Sustainable
Development Unit
Europe and Central Asia Region
1818 H Street, NW
Washington, DC 20433, USA

Tel: +1 202 473 62 29
Fax: +1 202 614 06 96
E-mail: sschwager@worldbank.org

Mr. Konrad VON RITTER
Lead Environmental Economist
Environmentally and Socially Sustainable
Development Dept.
1818 H Street, NW
Washington, DC 20433, USA

Tel: +1 202 458 04 77
Fax: +1 202 614 06 96
E-mail: kritter@worldbank.org

Mr. Anil MARKANDYA
Environmentally and Socially Sustainable
Development Dept.
1818 H Street, NW
Washington, DC 20433, USA

Tel: +1 202 423 21 89
Fax: +1 202 614 06 96
E-mail: amarkandya@
worldbank.org

Mr. Rune CASTBERG
Senior Environmental Specialist
Environmentally and Socially Sustainable
Development Dept.
1818 H Street, NW
Washington, DC 20433, USA

Tel: +1 202 473 86 31
Fax: +1 202 614 09 62
E-mail: rcastberg@worldbank.org

Mr. Walter STOTMANN
The World Bank
1818 H Street, NW
Washington, DC 20433, USA

Tel: +1 202 473 24 95
E-mail: wstottmann@worldbank.org

	Ms. Amy EVANS	Tel: +1 202 473 80 18
	Caspian Environment Program Team Leader	Fax: +1 202 614 07 09
	Environmentally and Socially Sustainable	E-mail: aevans@worldbank.org
	Development Dept.	
	1818 H Street, NW	
	Washington, DC 20433, USA	
IFC	Mr. William BULMER	Tel: +1 202 473 87 50
	Sector Manager	Fax: +1 202 974 43 18/9
	Infrastructure Department	E-mail: wbulmer@ifc.org
	International Finance Corporation (IFC)	Web-site: www.ifc.org
	2121 Pennsylvania Avenue, NW	
	Washington, DC 20433, USA	
EBRD	Mr. Joachim JAHNKE	Tel: +44 20 7338 7498
	Vice-President, ESE Vice-Presidency	Fax: +44 20 7338 6998
	European Bank for Reconstruction	E-mail: bryantl@ebrd.com
	and Development	
	One Exchange Square	
	London EC2A 2JN, UK	
	Mr. Timothy MURPHY	Tel: +44 20 7338 6020
	Director of the Environmental Appraisal Unit	Fax: +44 20 7338 6848
	EBRD	E-mail: murphyt@ebrd.com
	One Exchange Square	
	London EC2A 2JN, UK	
	Mr. Thomas MAIER	Tel: +44 20 7338 7924
	Acting Director of the Municipal	Fax: +44 20 7338 6964
	and Environmental Infrastructure Dept.,	E-mail: maiert@ebrd.com
	EBRD One Exchange Square	
	London EC2A 2JN, UK	
	Mr. Nils Christian HOLM	Tel: +44 20 7338 6546
	Principal Banker	Fax: +44 20 7338 6964
	Municipal and Environmental Infrastructure	E-mail: holmn@ebrd.com
	Development Dept., EBRD	
	One Exchange Square	
	London EC2A 2JN, UK	
	Mr. Ulf HINDSTROM	
	EBRD	
	One Exchange Square	
	London EC2A 2JN, UK	
PPC	Mr. Laurent GUYE	Tel: +4131 324 0829
	Minister	Fax: +4131 324 0954
	Federal Department for Economic Affairs	E-mail: laurent.guye@seco.admin.ch
	State Secretariat for Economic Affairs,	
	Chairman of the PPC	
	Effingerstrasse, 1	
	CH-3003, Bern	
	Switzerland	
	Mr. Anders RISBERG	Tel: +44 20 7338 7124
	Project Preparation Committee Secretary	Fax: +44 20 7338 6848
	EBRD	E-mail: Risberga@ebrd.com
	One Exchange Square	
	London EC2A 2JN, UK	
	Ms. Nathalie ROTH	Tel: +44 20 7338 7568
	PPC Officer, Power and Energy Utilities,	Fax: +44 20 7338 7280
	EBRD	E-mail: rothn@ebrd.com
	One Exchange Square	
	London EC2A 2JN, UK	
	Ms. Nadine WARREN	Tel: +44 20 7338 6327
	Administrative Secretary	Fax: +44 20 7338 6848
	Project Preparation Committee, EBRD	E-mail: Warrenn@ebrd.com
	One Exchange Square	
	London EC2A 2JN, UK	

Annex 2

ASIAN DEVELOPMENT BANK	**Mr. Jyrki WARTIOVAARA** Environment Specialist, Environment Division Office of Environment and Social Development Asian Development Bank 6 ADB Avenue, Mandaluyong City P.O. Box 789, 0980 Manila, Philippins	Tel: +632 632 6720 / 5333 Fax: +632 632 2195 E-mail: jwartiovaara@adb.org
UNECE	**Mr. Branko BOSNJAKOVICH** Regional Adviser of Environment United Nations Economic Commission for Europe Palais des Nations – Room 327 8-14, Avenue de la Paix CH-1211 Geneva 10 Switzerland	Tel: +41 22 917 2396 / 917 2361 Fax: +41 22 917 0107 / 917 0036 E-mail: branko.bosnjakovic@unece.org
UNEP	**Mr. Frederick SCHLINGEMANN** Director and Regional Representative of the Regional Office for Europe of the UNEP 15, Chemin des Anémones CH 1219 Chatelaine-Geneva Switzerland	Tel: +41 22 917 82 91 Fax: +41 22 797 34 20 E-mail: frits.schlingemann@unep.ch
	Mr. Merab SHARABIDZE Environmental Affairs Officer Regional Office for Europe, UNEP 15, Chemin des Anémones CH 1219 Chatelaine-Geneva Switzerland	Tel: +41 22 917 83 95 Fax: +41 22 797 80 24 E-mail: merab.sharabidze@unep.ch
UNEP/GRID	**Mr. Philippe REKACEWICZ** Visual Communication Unit UNEP GRID-Arendal Longum Park Service P.O. Box 706 Norway	Tel: +47 370 35 650 Fax: +47 370 35 050, 47 928 15 483 E-mail: reka@grida.no
	Ms. Ieva RUCHEVSKA Environmental Reporting and Indicators UNEP GRID-Arendal Longum Park Service P.O. Box 706 Norway	Tel: +47 370 35 650 Fax: +47 370 35 050 E-mail: ieva@grida.no
UNDP	**Mr. Fikret AKCURA** UNDP Officet-in-Charge United Nations Development Program 67, Tole bi Str. Almaty, Kazakhstan	Tel: +7 3272 58 26 37 Fax: +7 3272 58 26 45 E-mail: fikret.akcura@undp.org
	Mr. Selvakumaran RAMACHANDRAN Deputy Resident Representative United Nations Development Program 67, Tole bi Str. Almaty, Kazakhstan	Tel: +7 3272 58 26 37 Fax: +7 3272 58 26 45 E-mail: selva.ramachandran@undp.org
	Mr. Zharas TAKENOV Regional Specialist for Sustainable Development Policy, UNDP 67, Tole bi Str. Almaty, Kazakhstan	Tel: +7 3272 58 26 37 Fax: +7 3272 58 26 45 E-mail: Zharas.takenov@undp.org
	Mr. Andrey DEMYDENKO Project Manager, UNDP Aral Sea Basin Capacity Development Regional Project for Central Asia 4, Taras Shevchenko Street Tashkent 700029, Uzbekistan	Tel: +998 712 41 48 11 Fax: +998 71 120 67 18 E-mail: aod@aral.uz, ademydenko@aral-sea.net

	Mr. Christopher BRIGGS Regional Co-ordinator Regional Bureau for Europe and the CIS (RBEC), Global Environmental Facility (GEF), UNDP One UN Plaza, Room DC1-1668 New York, NY 10017, USA	Tel: +1 212 906 50 60 Fax: +1 212 906 51 02 E-mail: Christopher.briggs@undp.org
UNICEF	**Mr. Thomas THOMSEN** Area Representative Central Asian Republics and Kazakhstan UNICEF Area Office for Central Asian Republics and Kazakhstan 15, Republican Square Almaty 480013 Kazakhstan	Tel: +7 3272 50 16 65 Fax: +7 3272 50 16 62 E-mail: tthomsen@unicef.org
	Mr. Brendan A. DOYLE UNICEF Consultant 1069 Egan Avenue Pacific Grove, CA 93950 USA	Tel: +1 831 375 44 96 Fax: +1 831 649 02 72 E-mail: DoyleBrendan@aol.com
WHO	**Mr. Bent Hauch FENGER** Co-ordinator of the Health and Sustainable Development Section Regional Office for Europe World Health Organization 8, Scherfigsvei DK 2100, Copenhagen Denmark	Tel: +45 3917 1289 Fax: +45 3917 1890 E-mail: bfe@who.dk
	Mr. Kubanychbek MONOLBAEV Head of the Secretariat NEHAP Co-ordinator for NIS WHO Secretariat for NEHAP Implementation in Central Asian Republics Ul. Druzhby Narodov 46, 7th floor 700097 Tashkent Uzbekistan	Tel: +998 711 73 55 75 Fax: +998 712 77 60 35 E-mail: mkm@carnehap.uz
PRIVATE SECTOR	**Ms. Mariana ITEVA** Market Development Severn Trent Water International Ltd. World Business Council for Business and Environment 2308 Coventry Road Birmingham B26 3JZ United Kingdom	Tel: +44 779 88 13 394 Fax: +44 870 16 45 005 E-mail: Mariana.iteva@btinternet.com
	Mr. Jean-Patrice POIRIER Director for South-Eastern Europe and Central Asia 2 Vivendi Water 52, rue d'Anjou Paris 75384, Paris CEDEX	Tel: +33 1 49 24 33 53 Fax: +33 1 49 24 61 10 E-mail: jean-patrice.poirier@ generale-des-eaux.net
REC CEE	**Ms. Oreola IVANOVA** Head, Environmental Policy Programme Regional Environmental Center for CEE Ady Endre u.9-11 2000 Szentendre Hungary	Tel: +36 26 311 199 Fax: +36 26 311 294 E-mail: oivanova@rec.org
	Mr. Stefan SPECK Project Manager, Economic Instruments Environment Policy Programme Regional Environmental Center for CEE Ady Endre u.9-11 2000 Szentendre Hungary	Tel: +36 26 311 199 Fax: +36 26 311 294 E-mail: sspeck@rec.org

Annex 2

REC MOLDOVA	Mr. Victor COTRUTA Executive Director REC for Moldova 57/1 Banulescu Bodoni Str. Office 404 Chisinau, MD-2005 Moldavia	Tel: +373 2 23 86 85 Fax: +373 2 23 86 86 E-mail: Vcotruta@moldova.md
REC RUSSIA	Mr. Michael KOZELTSEV Executive Director REC for Russia 42, Slavayanskaya Sq., r.211 Moscow, Russia	Tel: +7 095 924 62 40 Fax: +7 095 925 92 82 E-mail: michael-rrec@mtu-net.ru
REC CENTRAL ASIA	Mr. Bulat YESSEKIN Executive Director REC for Central Asia 85, Dostyk Ave. Almaty, Kazakhstan	Tel: +7 3272 980 538 Fax: +7 3272 507 784 E-mail: Besekin@neapsd.kz
ENVIRONMENTAL NGOs	Mr. Sestager AKNAZAROV Director of NGO "Ecology of Biosphere" 95a Karasay Batyra Str., office 213 480012 Almaty, Kazakhstan	Tel: +7 3272 92 37 31 Fax: +7 3272 92 37 31 E-mail: aknaz@nursat.kz
	Ms. Olga PONIZOVA Co-ordinator European ECO-Forum/ECO-Accord 37-61, Sheremetyevskaya Str. 127521 Moscow Russian Federation	Tel: +7 095 924 62 40 Fax: +7 095 925 92 82 E-mail: accord@olgapon.gins.msk.su
	Ms. Anna GOLUBOVSKA-ONISIMOVA Director of NGO "MAMA-86" 22, Mikhailovska Str. 01001 Kiev-1 Ukraine	Tel: +380 44 2287749/2283101/ 229 55 14 Fax: +380 44 229 55 14 E-mail: anna@gluk.org Web-site: www.mama-86.kiev.ua
	Mr. Andrei ARANBAEV NGO, Turkmenistan	
	Mr. Yousup KAMALOV Chairman of the Union of Protection of Aral Sea and Amudarya 41, Berdakha Ave. Nukus, Karakalpakstan 742000 Uzbekistan	Tel: +998 361 21 77 229 E-mail: udasa@uzpak.uz
	Mr. Grant SARKISYAN Director of the Eco-club "Tapan" Kvartal ?-2, house 21, Apt. 23 Iugo-zapadnyi massiv 375114 Yerevan Armenia	Tel: +374 1 56 60 16 Fax: +374 1 56 60 16 E-mail: grant@ tapan.infocom.amilink.net
EAP TASK FORCE	Mr. Arcadiy CAPCELEA EAP Task Force Co-Chair	Tel: +3732 22 62 54 Fax: +3732 22 07 48 E-mail: acapcelea@worldbank.org
OECD EAP TASK FORCE SECRETARIAT	Mr. Brendan GILLESPIE Head Non-Member Countries Branch Environment Directorate, OECD 2, rue André Pascal 75775 Paris Cedex 16 France	Tel: +331 45 24 93 02 Fax: +331 45 24 96 71 E-mail: brendan.gillespie@oecd.org

Ms. Eija KIISKINEN
Administrator
Non-Member Countries Branch
Environment Directorate, OECD
2, rue André Pascal
75775 Paris Cedex 16
France

Tel: +331 45 24 18 40
Fax: +331 45 24 96 71
E-mail: eija.kiiskinen@oecd.org

Ms. Anne CARIOU
Administrative Assistant
Non-Member Countries Branch
Environment Directorate, OECD
2, rue André Pascal
75775 Paris Cedex 16
France

Tel: +331 45 24 87 21
Fax: +331 45 24 96 71
E-mail: anne.cariou@oecd.org

ENVIRONMENT POLICY TEAM:

Mr. Krzysztof MICHALAK
Administrator
Environment Policy Team Co-ordinator
Non-Member Countries Branch
Environment Directorate, OECD
2, rue André Pascal
75775 Paris Cedex 16
France

Tel: +331 45 24 96 00
Fax: +331 45 24 96 71
E-mail: krzysztof.michalak@oecd.org

Ms. Olga SAVRAN
Project Manager
Non-Member Countries Branch
Environment Directorate, OECD
2, rue André Pascal
75775 Paris Cedex 16
France

Tel: +331 45 24 13 81
Fax: +331 45 24 96 71
E-mail: olga.savran@oecd.org

Ms. Angela BULARGA
Project Manager
Non-Member Countries Branch
Environment Directorate, OECD
2, rue André Pascal
75775 Paris Cedex 16
France

Tel: +331 45 24 98 63
Fax: +331 45 24 96 71
E-mail:
angela.bularga@oecd.org

ENVIRONMENT FINANCING TEAM:

Mr. Grzegorz PESZKO
Administrator
Environment Financing Team Co-ordinator
Non-Member Countries Branch
Environment Directorate, OECD
2, rue André Pascal
75775 Paris Cedex 16
France

Tel: +331 45 24 19 47
Fax: +331 45 24 96 71
E-mail: grzegorz.peszko@oecd.org

Ms. Nelly PETKOVA
Project Manager
Non-Member Countries Branch
Environment Directorate, OECD
2, rue André Pascal
75775 Paris Cedex 16
France

Tel: +331 45 24 17 66
Fax: +331 45 24 96 71
E-mail: nelly.petkova@oecd.org

Ms. Anna TCHELIAPOVA
Administrative Co-ordinator
Non-Member Countries Branch
Environment Directorate, OECD
2, rue André Pascal
75775 Paris Cedex 16
France

Tel: +331 45 24 81 85
Fax: +331 45 24 96 71
E-mail: anna.tcheliapova@oecd.org

ENVIRONMENTAL MANAGEMENT IN
ENTERPRISES (EMES) TEAM:

Mr. Peter BÖRKEY
Team Co-ordinator
Non-Member Countries Branch
Environment Directorate, OECD
2, rue André-Pascal
75775 Paris Cedex 16
France

Tel! +331 45 24 13 85
Fax: +331 45 24 96 71
E-mail: peter.borkey@oecd.org

INTERPRETERS

Mr. Alexander RESHETOV
Interpreter
5, Parnikovaya Str., Apt. 72, Minsk

Tel: +375 172 211 88 56
E-mail: reshetov@minsk.sovam.com

Mr. Edouard PROKOPOVITCH
Interpreter
26, K. Marks Str., Apt.11, Minsk

Tel: +375 172 227 83 16
E-mail: reshetov@minsk.sovam.com

Mr. Igor ROMANENKO,
Interpreter
Almaty, Kazakhstan

Ms. Shafira KALDYBEKOVA
Interpreter
Almaty, Kazakhstan

Ms. Tatyana ZHAKUPOVA
Interpreter
Almaty, Kazakhstan

OECD PUBLICATIONS, 2, rue André-Pascal, 75775 PARIS CEDEX 16
PRINTED IN FRANCE
(97 2001 08 1 P) ISBN 92-64-18701-4 – No. 51889 2001